Worldwide Advances in Communication Networks

Worldwide Advances in Communication Networks

Edited by

Bijan Jabbari
Telecommunications Laboratory
George Mason University
Fairfax, Virginia

Plenum Press • New York and London

Library of Congress Cataloging-in-Publication Data

Worldwide advances in communication networks / edited by Bijan
  Jabbari.
        p.   cm.
    "Proceedings of an IEEE Symposium on Worldwide Advances in
  Communication Networks, held May 14-15, 1992, in Fairfax, Virginia"-
  -T.p. verso.
    Includes bibliographical references and index.
    ISBN 0-306-44818-1
    1. Telecommunication--Technological innovations--Congresses.
  2. Computer networks--Technological innovations--Congresses.
  I. Jabbari, Bijan. II. Institute of Electrical and Electronics
  Engineers. III. IEEE Symposium on Worldwide Advances in
  Communication Networks (1992 : Fairfax, Va.)
  TK5101.A1W69 1994
  621.382--dc20   94-24462
                                                                  CIP

Proceedings of an IEEE Symposium on Worldwide Advances in Communication Networks,
held May 14–15, 1992, in Fairfax, Virginia

ISBN 0-306-44818-1

©1994 Plenum Press, New York
A Division of Plenum Publishing Corporation
233 Spring Street, New York, N.Y. 10013

Printed in the United States of America

Foreword

The symposium on "Worldwide Advances in Communications Networks" which was held on May 14-15, 1992 at GMU was an ambitious attempt to bring together leaders in the communications area to discuss the major issues in this rapidly-changing technology. The symposium was a success and many of the ideas presented at the conference are being implemented.

This proceeding contains the majority of the papers presented at the symposium and abstracts of the remainder. The papers may be divided into seven general categories.

The first five papers explore some important design issues for high speed networks (gigabit networks). Traffic modelling, quality of service guarantees, switching alternatives, and routing are discussed.

The next two papers focus on applications for broadband communications. Weinstein begins by asking, "ARE THERE ANY APPLICATIONS?" and then proceeds to develop a wide variety of potential uses. Personick concentrates on multimedia applications.

The next three papers deal with Personal Communications Services (PCS) and the notion of communicating with anyone, at any time, anywhere. Several of the key technical issues such as CDMA vs TDMA are analyzed in detail.

The fourth area is satellite communications. Two papers discuss some of the major changes that are taking place and potential new systems.

The next two papers discuss signal coding and digital video. Jayant provides an excellent overview of the impressive capabilities that are available for the compression of speech, audio, image, and video signals. Bellisio concentrates on video encoding.

The next two papers deal with optics. The first discusses free-space digital optics as a potential hardware platform for digital signals. The second discusses issues concerning the Fiber Distributed Data Interface (FDDI).

The final two papers address security issues. The first develops routing protocols that are resistant to sabotage, and the second discusses privacy.

The overall Proceedings does an excellent job of presenting a wide array of crucial topics in communications networks. The organizing committee, chaired by Professor Bijan Jabbari of GMU, should be congratulated on this achievement.

Harry L. Van Trees
Distinguished Professor of Information Technology
Electrical and Systems Engineering

Preface

When the symposium was about to take place, I had numerous requests asking me to consider to put together a symposium proceeding reflecting some details of their presentations. Although, at the time of symposium preparation, we did not have any intent to have a proceedings, I approached the presenters with the idea of writing about the specific topic that they had given the presentation on. Despite their busy schedule, the request was taken favorably by many of the speakers and the response was encouraging. The current volume presents selected and updated version of the write-ups reflecting the presentations in this symposium.

The opening of the symposium consisted of three distinct presentations: The keynote talk presented by Paul Baran, the program overview by Harry Van Trees and the opening technical talk by Robert Gallager. These three presentations truly set the stage for an unforgettable event.

In his keynote talk on the impact of the evolving communications technology, Paul Baran emphasized the orders of magnitude improvements in our communications capability over the next decade and posed the following question: "How can society best use this major new potential resource and avoid some of the negative consequences that may be expected if we are not careful?" He stated that our move to information society requires some fundamental change in thinking. A greatly increased true education for students, as well as constant retraining of the labor force, are mandatory and communications technology can help achieve these programs for less than what we spend today.

I would like to thank all those presenters in this symposium. Special thanks go to the authors of the papers in this volume, to the organizing committee, and all those who made this possible. I am thankful to all my students in the Telecommunication Laboratory at George Mason University whose enthusiasm and generous help made this symposium a reality. In particular, I would like to thank Sirin Tekinay, Usha Narayanan, Yu Kuo, and Ferit Yegenoglu. I would like to thank Professor Harry Van Trees, Professor Gerald Cook, and Professor Raymond Pickholtz for their generous support. I am grateful to Mrs. Geet Sachidananda, who helped in preparation of the manuscript.

Bijan Jabbari
Faifax, Virginia

CONTENTS

Traffic Modelling for Broadband Services 1
    David M. Lucantoni, Marcel F. Neuts
    and Amy R. Reibman

Video and Multimedia Transport over Packet Media 9
    D. Raychaudhuri

Providing Quality of Service Guarantees in High-Speed Networks 19
    Jim Kurose

High Speed Switching Alternatives for Broadband Communications 27
    Maurizio Decina

Design, Dimensioning, Routing, and Control of Gigabit Networks 33
    Joseph Y. Hui

Applications for Broadband Communications 39
    Stephen B. Weinstein

Multimedia Applications and Their Communications Needs 51
    Stewart D. Personick

Applied Microcell Technology in the PCS Environment 57
    William C. Y. Lee

Issues in Microcellular Communications- CDMA Versus TDMA 65
    Raymond L. Pickholtz and Branimir R. Vojcic

The Millicom/SCS Mobilecom PCN Field Test 85
    Donald L. Schilling, Laurence B. Milstein,
    Raymond L. Pickholtz, Marvin Kullback and
    William Biederman

Control Architectures for Fixed and Mobile Networks 91
    Bijan Jabbari and Sirin Tekinay

Mobile-User Networking: The Satellite Alternative 99
    Anthony Ephremides and Jeffrey E. Wieselthier

Advances in Satellite Networks 107
    William W. Wu

Signal Coding: The Next Decade 125
    Nikil Jayant

Current Trends in Digital Video 137
    Jules A. Bellisio

Progress in Free-Space Digital Optics 143
    H. Scott Hinton

FDDI: Current Issues and Future Plans 149
    Raj Jain

Routing Protocols Resistant to Sabotage 167
    Radia Perlman

Internet Infastructure for Privacy-Enhanced Mail 171
    Robert W. Shirley

Abstracts 177

Index 185

TRAFFIC MODELLING FOR BROADBAND SERVICES

David M. Lucantoni[1], Marcel F. Neuts[2] and Amy R. Reibman[1]

ABSTRACT

There are many models being proposed for predicting the performance of multiplexed variable bit rate video sources. While these are important for engineering a network, it is clear that models of a single source are important for parameter negotiations and call admittance algorithms. In this paper we propose two novel performance measures which may be used to assess the "goodness-of-fit" of a model.

The first goodness-of-fit measure is a leaky bucket contour plot and may be used to quantify the burstiness of any type of traffic. A second measure is also proposed which applies only to video traffic and is based on the quality of the video predicted by the model.

INTRODUCTION

It is well recognized that the viability of B-ISDN/ATM depends on the development of effective and implementable congestion control schemes. While many frameworks and techniques are under discussion (See e.g., [1]), at least two capabilities have been agreed to as necessary in any framework that might arise. The first is a connection admission control (CAC) by which the network will decide to accept or reject a new connection based on a set of agreed to traffic descriptors and on available resources. Once a connection is accepted, a second necessary control is some form of usage parameter control (UPC) which will insure that connections stay within their negotiated resource parameters. A popular UPC would involve a leaky bucket monitor of traffic entering the system, where traffic deemed as excessive by the monitor could either be dropped or

[1]AT&T Bell Laboratories, Holmdel NJ 07733
[2]Dept. of Systems and Industrial Engineering, University of Arizona, Tucson,AZ 85721

tagged as low priority and allowed to proceed through the network to take advantage of potentially unused resources.

Performance modeling is necessary to determine which techniques or set of techniques will be appropriate for eventual implementation in a B-ISDN network. Such models need to take into account traffic characteristics from realistic services that would be carried in a B-ISDN network. In particular, we need traffic models which will accurately represent the statistical nature of very high-speed, bursty services.

We also need metrics to quantitatively assess when a model is good. Standard statistical measures such as means, variances, and other goodness-of-fit tests may not be appropriate here since they may not be measuring the characteristics of the process that are most important for either predicting the effect of the source on the resources in the network or the performance the source will experience. Instead, we would like measures that are related to what will affect the performance that the source experiences. Such measures can also be used in comparing different models of the same source.

In this paper, we propose two novel performance measures for quantitatively evaluating the goodness-of-fit of models. In the next section we specify the goals that we want to achieve with a traffic model. A new measure of traffic burstiness is proposed and a measure of video quality for VBR video is also discussed.

GOALS OF TRAFFIC MODELS

Two classes of traffic models need to be developed: multiplexed source models and single source models. Although the same traffic model might be used in both cases, some models might be more suitable for one than the other. Multiplexed models will capture the effects of statistically multiplexing bursty sources and will predict to what extent the superposition of bursty streams is "smoothed". These models will be useful in traffic engineering the network (e.g., deciding how many links or virtual paths to put between different locations) and in traffic management (e.g., designing connection admission control algorithms, etc.) Several models have already been proposed in this direction (see, e.g., [2], [3], [4], [5] and the references there).

There are several areas where single source models are useful. They could be used to study what types of traffic descriptors make sense for parameter negotiation between the network and the end system at call setup. For example, if leaky bucket monitoring is used as a traffic descriptor, the negotiation might consist of the source specifying what parameters could be used in the leaky bucket for a given connection. Single source models of different applications would help in the selection of these parameters. Also, some applications might want to do some kind of end-to-end rate control to ensure that minimal traffic is lost during periods of network congestion. Source models could be used in testing various rate control algorithms. Finally, these models are also useful in predicting the quality-of-service (QOS) that a particular application might experience during different levels of congestion.

In deriving traffic models, we need metrics which can determine how "close" the model is to the actual traffic. Standard statistical "goodness-of-fit" tests are inappropriate here since it is not necessary that the traffic generated by the model "look" like the real traffic. However, enough of its important characteristics should be matched so that specific performance predictions are accurate. That is, the goodness-of-fit metrics need to be directly related to the specific aspects of performance that we want to predict from the model. (See e.g., [6])

In this paper, we propose two criteria for judging the appropriateness of a traffic model for bursty services. The first one applies to any high speed bursty data service and the second is specific to a variable-bit-rate (VBR) video application. We compare a previous model of VBR video with a new Markov renewal process model proposed in [7].

A MEASURE OF TRAFFIC BURSTINESS

A diagram of a generic video system is shown in Figure 1. Note that between the video encoder (which generates the bits per frame) and the transmission channel we have indicated a buffer/bucket. This could represent a physical buffer that contains the actual traffic and performs some type of traffic shaping, or it could represent a logical buffer such as a leaky bucket counter[8] which only monitors and does not buffer the traffic. A physical buffer is currently used in all CBR video systems to ensure that the bit rate onto the channel is constant. (See e.g., [9]) A logical buffer could be used by the VBR system. In either case the end user would like to know the parameters of the buffer or bucket that are needed to ensure that a given percentage of cells are excessive. The parameters that are needed are directly related to the average rate and burstiness of the source. Since a physical buffer is logically equivalent to a leaky bucket with a given drain rate, we will focus our attention on leaky buckets for the remainder of this discussion.

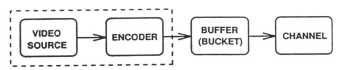

Figure 1. A Video System.

It has been proposed previously (see, e.g., [8]) that the leaky bucket parameters (e.g., a given bucket size and drain rate) that a source would want the network to use in a monitor could be passed to the network as a descriptor of the traffic characteristics. While we still believe this to be a valid approach in terms of getting useful traffic characteristics from a simple small set of parameters, it is clear that it is only a partial characterization of the traffic. To judge the goodness-of-fit of a model or to compare different models we would like to have a more comprehensive characterization of the traffic. We propose the concept of leaky bucket contours as a way to describe the traffic characteristics of a bursty data source. (Note that this concept is not restricted to video traffic and in fact can be used to describe any bursty data source.)

Typical plots of leaky bucket contours are shown in Figure 2. The drain rate of the leaky bucket is plotted on the x-axis and the size of the bucket is plotted on the y-axis. Points along the different curves are those pairs of (drain rate, bucket size) which result in a fixed percentage of traffic overflowing a leaky bucket with the given parameters. If we think of representing the probability of overflow as the z-axis coming out of the page, then the curves that are shown can be viewed as contour lines of a 3-dimensional surface which is of height 1 at the origin of the x-y axes and which decreases in height as we get further into the upper right hand quadrant. The steepness with which this surface approaches zero on the z-axis is directly related to the burstiness of the traffic. The contour plots shown in Figure 2a are clearly from a less bursty source than that shown in 2b. The leaky bucket contours (or at least one contour) may also help the end system decide which parameter values to choose for negotiation with the network.

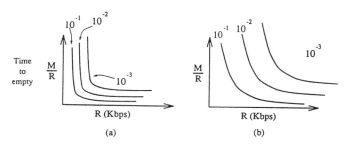

Figure 2. A Characterization of Traffic Burstiness.

One reason that there has never been a satisfactory definition of burstiness which could be standardized is because when we talk about burstiness we need to mention what time scales we are measuring the burstiness on. These leaky bucket contours capture the burstiness on many time scales simultaneously. A very good feeling for the burstiness of the source can be obtained by observing the gradients along the surface at these higher probability contours. If we needed to give a measure of the burstiness of a data traffic stream in terms of just one parameter, the volume in the positive quadrant of the surface described by the contours might be an appropriate measure. Clearly, the larger the volume, the more bursty the source for the same actual (average) rate.

Several comments are in order. Note that we have shown typical contours corresponding to height of 10^{-i}, $i = 1,2,3$. At first this may not seem consistent with the view that typical leaky bucket parameters might be chosen to result in overflow probabilities of 10^{-6} - 10^{-9} or less. We believe, however, that the higher probability contours are relevant for several reasons. First, these are the only contours that can be accurately estimated with the limited amounts of data available. For example, in the VBR video data, although we have 9000 samples we see in [7] that we still don't get a very accurate estimate of the 10^{-2} curve. The reason for this is that these are not independent samples but are in fact highly correlated. Therefore, there is no way that the 10^{-6} or lower curves can be estimated to a high degree of accuracy for any reasonable simulation run.

Another argument for the relevance of these contours is the following. If a data source really wanted to ensure that 10^{-9} percentage of its traffic would overflow a leaky bucket then there are two possibilities. The first is to set very large drain rate and/or bucket sizes for the network to monitor the traffic. This would result in a large amount of resources (i.e., buffers, bandwidth, etc.) that the network would in effect have to reserve for the source, which would translate into a very high cost to the end user. The second possibility which we believe to be the more likely one for most sources, is to set more reasonable leaky bucket parameters to reduce the cost of transport and to do at least a minimal amount of traffic shaping to ensure that no more than the desired percentage of traffic would overflow the bucket. Most sources would probably set the parameters in the 10^{-1} - 10^{-3} range and shape the traffic when necessary. The contour plots in this range then give the end user a very good idea of how much shaping will be necessary for different sets of parameters.

A HEURISTIC METRIC OF VIDEO QUALITY

The leaky bucket contours discussed in the last section can be used to compare the burstiness of different sources or to quantify the goodness-of-fit of a traffic model to a source. As mentioned earlier, this measure is appropriate for describing any bursty source. In this section we propose another measure which can be used to judge the merits of a model, however, this measure is only appropriate for packet video models. To describe this measure we first need a little background on the operation of a video coder with rate control.

Consider a video system displayed in Figure 3. The CCITT H.261 video standard requires that a constant bit rate (CBR) be offered to the channel. This can be achieved by employing an adaptive encoder rate control as follows. As bits are generated in the encoder they are placed in a buffer before being transmitted onto the channel. The buffer empties at a constant rate onto the channel. To ensure that the buffer does not overflow the content of the buffer is monitored in real time. If the buffer starts to fill, the rate control specifies that the encoder increase the quantization step size (q) used for the discrete cosine transform coefficients. This reduces the amount of information per frame and hence reduces the bit rate out of the encoder. If the buffer fullness decreases, the rate control decreases q which increases the bit rate out of the encoder. By picking the thresholds on the buffer content for adjusting the quantization size appropriately, we may ensure that the buffer never overflows or completely empties.

The above description of a CBR source holds for a VBR source if we assume that the network monitors the traffic using a leaky bucket with specified parameters and that the source does not want any traffic identified as excessive by the network. In this case, the output of the encoder is transmitted directly to the channel (i.e., it is truly VBR), but the source monitors the output by mimicking the leaky bucket monitoring of the network. To ensure that the offered traffic is compliant with the leaky bucket parameters, the video system again does rate control by monitoring the content of the leaky bucket and adjusting the quantization step size appropriately. We see that the rate control operation is identical when either a physical or a logical (e.g., leaky bucket monitor) buffer is being monitored.

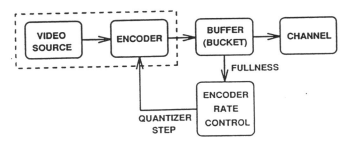

Figure 3. Video System with Rate Control.

Clearly, changing the quantization step size will have an impact on the quality of the video signal. That is, if most of the step sizes used are small, then more information is retained and a higher resolution video signal is transmitted. On the other hand, larger quantization implies poorer video quality. Therefore, monitoring the sequence of quantization step sizes used will give an indication of the quality of the transmitted video signal. It is also clear that the actual video quality depends on the correlation structure of the sequence of quantization steps. That is, a video sequence which consists of a long string of data encoded with a small step size and another long string encoded with a large step size would have an overall quality much different than a sequence with the same proportion of small and large steps which are randomly distributed in the sequence.

We propose as a first step in quantifying the quality of the VBR signal to ignore the serial correlations among the sequence of quantization sizes and just look at the marginal distribution of the quantization sizes. In general, comparing two histograms of the quantizations used for different video streams may not allow us to say that one stream has a higher quality than the other, but if a histogram is shifted towards smaller quantization sizes then we could claim that it represents a better quality signal. It is in this sense that we call this measure a heuristic.

In principle, we could directly model the joint process of the bit rate produced by and the q-step used by a video encoder with rate control. However, a complex model would be needed to capture accurately many of the variables involved e.g., the within-frame picture variations, the effect of memory when changing q, the effect of changing q for different image content, etc. Instead, in [7] we use an existing model of the bit rate produced by a video encoder with constant q-step, and approximate the gross effect of changing the q-step using a simple scaling model.

To be specific, we note that our original VBR video data was encoded using a fixed quantization step size of q = 8. (See e.g., [10]). This resulted in a given long term average rate for the source (call it r_8). Next, we re-encoded the entire stream with q = 10 and compute the average rate r_{10}. We continue this process for a number of values of the quantization step size, say q = 4,6,...,20, etc. From these rates we plot a scaling factor as a function of the step size as shown in Figure 4, where the step size q is on the x-axis and the scale factor r_q / r_8 is on the y-axis.

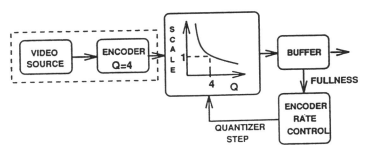

Figure 4. Approximation to Video System.

The effect of the rate control is approximated as follows. We start generating traffic according to either the original data or a traffic model that predicts the bit rate of a codec with q = 8. This traffic is monitored by a leaky bucket algorithm. Based on the content of the leaky bucket and a given set of rules for the rate control algorithm, a new quantization size, q, may be prescribed. This decision is made 10 times a frame. At this point we scale the output bits per frame by the scaling factor associated with the new q. The traffic with the new scaled rate continues to be monitored by the leaky bucket and the rate control.

By separating the effect of the different quantizer step sizes from the effect of the rest of the encoder, we considerably simplify the modeling process and any simulations that will be run. For example, without the separation, if either the channel rate or the rate control algorithm changed it would be necessary to recompute the joint model of the bit rate and the quantization step sizes before running any simulations. However, with the separation, the model will still be valid and previous simulation results could be reused. An illustration of the video quality heuristic is given in [7].

CONCLUSIONS

In this paper we propose two new performance measures. The first is a characterization of traffic burstiness which is directly related to what the stream will encounter in the network (i.e., buffers or buckets). This can be used to compare the burstiness of different traffic streams (not necessarily video) or to judge the goodness-of-fit of various traffic models. The second metric we propose is for comparing traffic models of VBR video by relating how well they predict the quality of the video call.

As mentioned above, the histogram of quantization sizes is only a first step towards predicting the actual quality of a video call. Further work is needed in testing whether models also capture the correlation in the sequence of quantization sizes and what effects these correlations have on the real subjective quality of a call. In fact, something that is currently missing from studies of video modeling is subjective testing to determine what types of realizations of quantization sizes are acceptable to real users, similar to tests that were done in determining the effect of bit-dropping in packetized voice.

REFERENCES

1. "Traffic Control and Congestion Control in B-ISDN," CCITT Recommendation I.371, Geneva, Switzerland, June, 1992.
2. Verbeist, W., Pinno, L., and Voeten, B., "The impact of the ATM concept on video coding", *IEEE J. Selected Areas Commun.*, Vol. SAC-6, No. 9, Dec., 1988.
3. Nomura, M., Fujii, T., and Ohta, N., "Basic characteristics of variable bit rate video coding in ATM environment". *IEEE J. Selected Areas Commun., Vol.* 7, pp. 752-760, June, 1989.
4. Maglaris, B., Anastassiou, D., Sen, P., and Roberts, J. D., "Performance models of statistical multiplexing in packet video communications", IEEE Trans. Commun., Vol. COM-36, No. 7, pp. 834 - 843, July, 1988.
5. Heyman, D., Tabatabai, A., and Lakshman, T.V., "Statistical analysis and simulation study of video teleconference traffic in ATM networks" IEEE Trans. on Circuits and Systems for Video Technology, Vol. 2. No. 1. March 1992.
6. Whitt, W., "Approximating a point process by a renewal process: the view through a queue - an indirect approach, Manage. Sci. Vol. 27, No. 6, pp.619-36, June, 1981.
7. Lucantoni, D. M., Neuts, M. F. and Reibman, A. R., Traffic models for variable-bit-rate video, submitted for publication.
8. Eckberg, A.E., Lucantoni, D.M., and Luan, D.T., An Approach to Controlling Congestion in ATM Networks, International Journal of Digital and Analog Communication Systems, Vol. 3, pp. 199-209, April/June 1990.
9. Netravali, A. N. and Haskell, B., Digital Pictures: Representation and Compression, Plenum Press, 1988.
10. Recommendations of the H-series, CCITT Study Group XV, Com XV-R 37-E, August 1990.

VIDEO AND MULTIMEDIA TRANSPORT OVER PACKET MEDIA

D. Raychaudhuri*

David Sarnoff Research Center
Princeton, NJ 08543-5300

ABSTRACT

This summary paper provides a brief overview of a 5/92 IEEE /GMU Symposium talk on current issues related to video and multimedia transport over networks. After an introductory discussion, design issues for key packet video/multimedia subsystems are considered. Topics covered include: (a) design and characterization of variable bit-rate (VBR) video compression algorithms; (b) prioritization/layering of video data for robustness; (c) protocols for reliable video transport; (d) cell-loss requirements for video-over-ATM.

INTRODUCTION

Emerging high-speed packet media (e.g., FDDI, SMDS, ATM/B-ISDN, Gigabit network..) represent a unique opportunity for the introduction of a new generation of high quality and cost-effective video/multimedia services. These networks provide wideband connectivity at a cost per bps substantially lower than the current communications infrastructure, thus enabling a variety of distributed multimedia applications. In addition to providing cost-effective broadband transport, packet/cell-relay networks offer qualitatively new services, such as selectable constant bit-rate (CBR) and variable bit-rate (VBR) video transport for multimedia terminals. The variable bit-rate (VBR) mode supported by the general class of packet-switched networks (including high-speed LAN's, WAN's and ATM) is of particular interest because of a number of potential technological benefits.

Transport of video and multimedia via packet media is associated with a range of near-, medium- and long-term application scenarios. Some example scenarios, along with required image quality, applicable image compression algorithms, bit-rate regime and transmission media are summarized in Table 1.

---

* Now with NEC USA, C & C Research Laboratories, Princeton, NJ 08540.

Table 1

Application	Image Qual. Required	Video Compr. Algorithms	Bit- Rate Regime	Applicable Transmission Media
Picturephone	Low	H.261,VQ, custom algs.	19.2-100Kbps	PSTN, narrowband ISDN, current cable LAN
1st generation multimedia desktop conf.	Med	H.261, MPEG -1, custom algs.	256Kbps- 1.5 Mbps	fiber LAN (e.g., FDDI), frame relay, telco SMDS
2nd generation multimedia	Med - High	MPEG -1, MPEG -2, CCITT/ATM, U.S.HDTV..	1.5 - 10 Mbps	high - speed fiber LAN (e.g., FDDI-II, ATM LAN,...) telco SMDS, ATM/B-ISDN, NREN/CNRI Gbps network
Broadcast TV	Med - High	MPEG -2, CCITT/ATM, U.S.HDTV	4 - 10 Mbps	custom cable & DBS, telco ATM/B-ISDN
HDTV	High	U.S. HDTV standard, CCITT/ATM, MPEG -2	10 - 50 Mbps	terrestrial broadcast, cable & DBS ATM/B-ISDN

The potential benefits of packet video for many of the above scenarios include: low latency/delay, service flexibility with true multimedia integration, lower transmission and switching cost due to resource sharing, constant image quality with VBR coding, effective support for multimedia applications with fast changing video/image/voice/data bit-rates, higher network capacity / bandwidth utilization (at a given video service quality), and potentially lower VBR codec cost/complexity.

KEY SUBSYSTEMS AND RELATED R&D TOPICS

The key subsystems of a packet video/multimedia system are [1]:

(1) Video codec suitable for network operation
(2) Video Network Interface Unit (NIU) for adaptation of codec to network
(3) Higher layer protocols for media synchronization, session control, presentation, etc.
(4) High-speed packet network with protocol appropriate for video/multimedia

Research issues related to the above subsystems are:

(1) Design of VBR compression algorithms for packet media, including consideration of: high/low priority layering for robustness under packet loss, multiresolution hierarchy for service flexibility, VBR encoder bit-rate properties and source models, end-to-end performance evaluation, etc.

(2) Design of transport protocols for video, including consideration of: transport/adaptation layer functionality for ATM (i.e. video AAL), segmentation and re-assembly requirements, time-stamping and delay compensation buffering, error control, flow control, priority support, etc.

(3) Network design/retrofit for improved real-time traffic support, including consideration of: medium access control (MAC) and service scheduling policies for time-constrained traffic, priority support, flow control and error control for real-time video/multimedia, etc.

The David Sarnoff Research Center has been active in a number of R&D areas outlined above, both in terms of design and prototype implementation. Recent projects include:

(1) Variable Bit-Rate (VBR) video algorithms for ATM/B-ISDN.
(2) Robust HDTV compression and transport for terrestrial, satellite, cable and ATM media.
(3) Distributed workstation-based multimedia via local/wide area high-speed networks.
(4) Variable bit-rate (VBR) H.261 compatible video over conventional local area networks.

The following provides a brief discussion of selected results from the above packet video/multimedia related projects.

SELECTED DISCUSSION TOPICS

Video Compression

Video compression algorithms for packet media may be:

- VBR (variable bit-rate) <u>or</u> CBR (constant bit-rate)
- one-layer <u>or</u> two-layer
- hierarchical <u>or</u> non-hierarchical
- standards-based <u>or</u> custom algorithm syntax

Selection from the above options will depend on a number of performance and technology factors including image quality needs, robustness/hierarchy requirements, compatibility with other services/systems, transmission cost, equipment cost, service evolution, product timing vs. VLSI availability, etc. Prominent approaches for video over

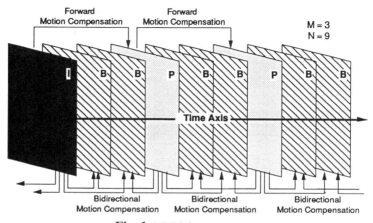

Fig. 1. MPEG frame structure

packet media under consideration include:

- standards based, 1-layer, non-hierarchical CBR (e.g., MPEG-1 standard, [2])
- standards based, 2-layer, CBR (e.g., MPEG++ proposed for HDTV, DBS & cable, [3])
- standards based, 1- or 2- layer CBR/VBR (e.g., MPEG-2 and CCITT SGXV ATM)
- custom, two-layer, hierarchical VBR for ATM (e.g., NTT ATM codec prototype, [4])

Standards based compression

MPEG: MPEG (motion picture experts group) is a DCT-based world standard for video compression, initially aimed at 1.2 Mbps CD applications, but is growing in acceptance & scope. MPEG-1 is based on a "Group-of-pictures" (GOP) structure with Intra (I) frames, Predicted (P) frames and Bidirectionally-interpolated (B) frames, as shown in Fig.1.

An example VBR MPEG-1 encoder is under preliminary consideration for network based multimedia applications, in view of the growing technological importance of the MPEG standard. The encoder follows MPEG-1 syntax, but without constant bit-rate control and rate buffering. The codec produces constant image quality by operating with a fixed set of quantizers. Where required, peak rate control is implemented via recoding of certain frames that exceed agreed-upon peak bit-rate constraints imposed by the network. An example of the VBR MPEG bit-rate process is shown in Figure 2 below [5]. Note the periodic structure of the bit-rate process due to the group-of-pictures I, B, P format shown in Fig. 1 above. This type of bit-rate process has interesting implications for resource-shared ATM network design; an analysis of statistical multiplexing performance will be reported in the near future.

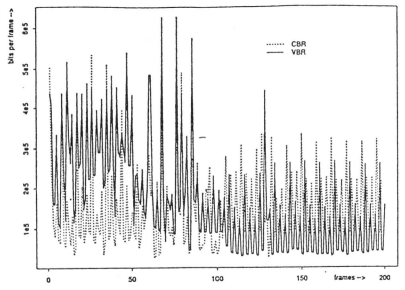

Fig. 2. Example VBR MPEG bit-rate process.

H.261: An alternative standards-based VBR compression approach is to use px64 Kbps H.261 (CCITT) syntax, but without rate control. VLSI for H.261 are currently available, and can be used to develop VBR packet video codecs. At the lower bit-rate levels (i.e., ~100-200 Kbps), the H.261 standard can be used for near-term workstation multimedia via conventional local area networks, [7]. Typical bit-rate histogram and autocorrelation plots for an example VBR H.261 encoder are shown in Fig. 3(a) and 3(b).

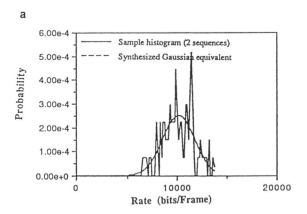

a

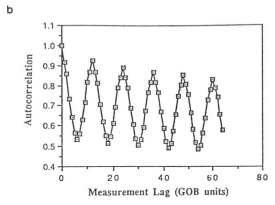

b

Fig. 3. Example VBR H.261 bit-rate process (a) histogram and (b) autocorrelation.

Video Layering / Prioritization for Robustness

Layering of video data into high and low priority bit-streams provides improved robustness in the presence of packet/ATM cell loss events [4]. The general principle employed has been to transmit subjectively important video information with high priority (HP), with the remaining less important information sent standard priority (SP). This approach provides "graceful degradation" in networks (such as ATM) with two priorities. Most recently, it has been used in the AD-HDTV terrestrial broadcasting system to provide robustness in the presence of interference and noise [6].
Two distinct approaches to compression algorithm layering have evolved:

1) external processing of standard one-layer codec's output.
(e.g., Sarnoff/Thomson/Philips/CLI's MPEG++)
2) custom hierarchical algorithm with embedded layering (e.g., future MPEG-2...).

The MPEG++ system proposed for HDTV compression operates at 17.72 Mbps with 20% data transmitted as "high priority" and 80% data transmitted as "standard priority". An example of simulated SNR results for MPEG++ layering is given in Fig. 4 below. The data shows that a moderate level of decoded image quality can be maintained even at very high SP loss rates. This feature has also been demonstrated in the AD-HDTV prototype hardware currently being tested by the FCC.

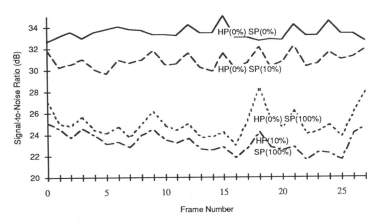

Fig. 4. Illustration of MPEG++ robustness due to two-tier layering.

It is noted that the ratio of high and standard priority information is a crucial design parameter, and will have to conform to the restrictions imposed by the medium under consideration. In ATM, some fraction (e.g., 10-30%) may be transmitted HP, but the additional robustness must be balanced against the increased transmission cost relative to one-tier. Two-tier layering is thus a useful technique in selected applications (e.g., broadcasting) with special robustness needs, but may not be appropriate in many communicative multimedia applications.

Video Transport

Video and multimedia systems based on packet media will require the definition and validation of a new generation of transport protocols. In particular, video as a data type requires somewhat different treatment than conventional data communication packets in the areas of segmentation/reassembly, flow control, error control, etc. Several candidate protocols have been proposed for emerging multimedia and TV/HDTV applications (e.g., ATM AAL, VMTP [7], NVP & PVP for Internet [8], MPEG++ [9],...).

Sarnoff's proposed MPEG++ transport approach for terrestrial, DBS (direct broadcast satellite) and cable is based on an ATM-type cell-relay transport structure with 3 distinct sublayers outlined in Table 2 below.

Table 2. MPEG++ Protocol Sublayers and Functions

Sublayer	Scope	Functions
Data-link/network	Generic and channel specific transport functions	Service multiplexing, HP/SP priority support, error detection(CRC), cell sequence indicator.
Video Adaptation	Service specific interface to data link/network sublayer	Segmentation and re-assembly of video data, alignment of HP/SP streams, logical video resynchronization after cell loss, error concealment support
Video service	Higher level service specific	Video decoder parameter and priority decoder reset

A basic issue in delivering video reliably over lossy cell relay networks is that of segmentation and reassembly (SAR) of variable length coded units of compressed data. For robust operation, a decoder with error concealment must be able to identify lost video blocks, and reenter at the appropriate resync point (e.g., MPEG "macroblock" or "slice"). This can be accomplished using the concept of an "entry pointer" in the video specific adaptation layer. The entry pointer principle is illustrated in Fig. 5 below; note that loss of cell #2 means that all of video unit B2 is lost, but that re-entry can occur at B3 in cell #3.

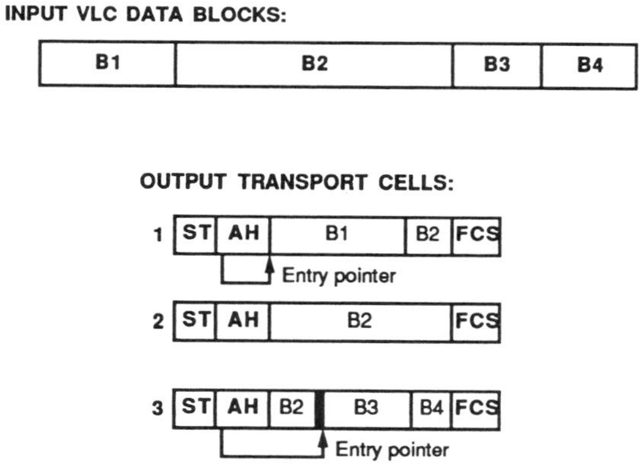

Fig. 5. The entry pointer principle for video segmentation & reassembly

The specific MPEG++ adaptation header implementation used in the AD-HDTV prototype is shown in Fig. 6 below. Note that in addition to the entry pointer, fields identifying video units and their key parameters are provided for error recovery.

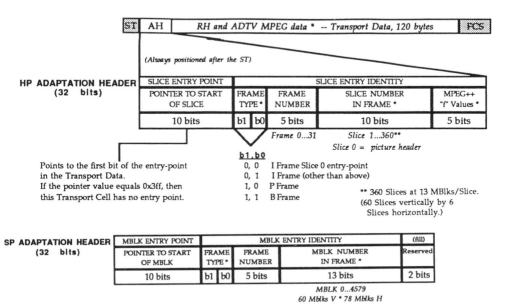

Fig. 6. MPEG++ adaptation layer

Video-over-ATM Demonstration

A tape demonstration of end-to-end MPEG++ transport over an ATM-type cell-relay network is available for viewing, [10]. The tape shows the results of a detailed simulation of 7 Mbps MPEG compressed video (at CCIR 601 resolution) transmitted using the MPEG++ transport protocol (128 byte payload + headers, etc.). Both one- and two-tier prioritized transmission are considered. Decoded image quality is shown at a range of cell loss rates between 10^{-5} and 10^{-1}. Our results show that with appropriate decoder error concealment, acceptable picture quality can be provided at cell loss rates as high as $\sim10^{-3}$ for one-tier transmission and $\sim10^{-2}$ for two-tier transmission. Results such as these have important implications on the design of ATM systems for video and multimedia - specifically, it appears that video can tolerate significantly higher cell loss rates than originally speculated. This suggests that VBR resource shared video networking is indeed feasible, and may be useful in application scenarios such as workstation-based distributed multimedia.

CONCLUSIONS

Next generation networks such as ATM/BISDN will undoubtedly be required to support a wide range of CBR and VBR video sources. Accordingly, understanding video transport is important since it could turn out to be the single largest source of traffic on many future high-speed packet networks (..and even on some lower speed ones...).

VBR offers many advantages over CBR, and should not be precluded by over-conservative network engineering principles. Resource-shared VBR operation is feasible on most networks when the average video bit-rate $\ll$ 10% of channel speed. Two-tier layering and prioritization of video may be helpful in such VBR scenarios, but is probably not essential. Standard one-layer video (e.g., MPEG, H.261..) can be supported on networks with moderate packet loss rate, provided that good transport protocols and concealment algorithms are used.

Overall, our results show that video is less "fragile" than widely believed due to its residual redundancy. For example, MPEG video with appropriate transport and error concealment requires an ATM cell loss rate $\sim 10^{-3}$ (not 10^{-7} - 10^{-9}) for reasonable quality. This has important implications for ongoing ATM network design activities. Real progress on an effective systems approach for video and multimedia over broadband networks will require active cooperation between the network & video communities

REFERENCES

[1] K. Joseph, D. Raychaudhuri and J. Zdepski, "Shared Access Packet Transmission systems for Compressed Digital Video", IEEE J. Selected Areas in Commun., June 1989, pp. 815-825.

[2] "MPEG Video Committee Draft," MPEG Video CD Editorial Committee, ISO-IEC JTC1/SC2/WG11 MPEG 90, December 18, 1990.

[3] Advanced Digital Television: Prototype Hardware Description, FCC Working Party 1 Final Certification Document, Feb. 1992.

[4] F. Kishino, et al, "Variable Bit-Rate Coding of Video Signals for ATM Networks", IEEE J. Selected Areas in Commun., June 1989, pp. 801-806.

[5] D. Reininger, J. Zdepski, D. Raychaudhuri, "VBR MPEG Bit-rate Characteristics", CCITT SGXV, Working Party XV/1, Experts Group for ATM Video Coding, Document AVC-307, July 1992.

[6] K. Joseph, et al, "Prioritization and Transport in the ADTV Digital Simulcast System", Proc. ICCE, June 1992.

[7] D. Cheriton, "VMTP: A transport protocol for the next generation of computer systems", Proc. ACM SIGCOMM '86 Symp., 1986.

[8] C. Topolcic, et al, "Experimental Internet Stream Protocol, Version 2 (ST-II)", RFC 1190, USC Information Sciences Institute, Oct. 1990.

[9] R. Siracusa, K. Joseph, J. Zdepski and D. Raychaudhuri, "Flexible and Robust Packet Transport for Digital HDTV", to appear in IEEE J. Selected Areas in Comm. (1992).

[10] R. Saint Girons, J. Zdepski and D. Raychaudhuri, "Transport and Error Concealment for MPEG-2", MPEG91/311, Kurihama, Nov. 1991.

PROVIDING QUALITY OF SERVICE GUARANTEES IN HIGH-SPEED NETWORKS

Jim Kurose

Department of Computer Science
University of Massachusetts
Amherst, MA 01003

ABSTRACT

In this paper we identify the issues involved in providing quality-of-service (QOS) guarantees to sessions in a high-speed network and briefly survey research in this area. Three approaches towards providing QOS guarantees are described and discussed: the tightly controlled approach, the approximate approach, and the bounding approach.

1. INTRODUCTION

Unlike traditional data networks, future broadband-ISDN (BISDN) wide-area networks will be required to carry a broad range of traffic classes ranging from bursty, variable-rate sources, such as voice and variable-rate coded video, to smooth, constant bit rate sources [22]. Moreover, these networks will have to do so while providing a guaranteed performance or quality-of-service (QOS) to many of these traffic classes. The ability to guarantee performance is particularly important in networks supporting real-time applications [20]. The actual QOS performance metrics of interest are likely to vary from one application to another, but are projected to include such measures as cell loss, delay, and delay jitter guarantees. The need to provide end-to-end QOS guarantees (essentially a circuit-like performance requirement) while still taking advantage of the resource gains offered by a statistically-multiplexed transport mechanism remains an important, yet largely unsolved problem facing BISDN architects.

The difficulties that arise in networks with QOS guarantees can be most easily illustrated by considering the "simple" question that must be answered each time a call/session arrives to such a network: "Can the requested call be accepted by the network at its requested QOS, without violating existing QOS guarantees made to on-going calls?" To answer this question, the network must be able to compute (or provide) not only the guaranteed end-to-end quality-of-service to be received by the arriving session, but also determine the performance impact of admitting this session on the already-accepted sessions in the network. The problem here then is one of performance-oriented call admission control [22]. We note that this problem is distinct from (but in some respects dependent upon) the lower-level controls (such as flow/congestion control and cell discarding policy) which are applied to the individual cells or bursts within the traffic streams of already-accepted calls. We refer to the call admission process as being "performance-oriented" since in order to answer the call admission question, the performance realized by an admitted session and its impact on the performance of existing sessions must be explicitly considered. Performance concerns, once traditionally relegated to the off-line tasks of dimensioning and design (e.g., determining link bandwidths, buffer capacities and processing capacities at network switchpoints) now become an on-line concern given this need for performance-driven traffic control mechanisms.

The introduction of QOS requirements into high-speed wide-area networks poses a large number of technical challenges. In this paper we identify the issues involved in providing QOS guarantees and briefly survey research aimed at resolving these challenges. The remainder of this paper is structured as follows. In section 2, we identify several different approaches towards defining QOS and the fundamental challenges in providing performance guarantees. In section 3, we overview three basic approaches for providing QOS guarantees. Section 4 concludes this paper.

2. QOS GUARANTEES: DEFINITIONS AND FUNDAMENTAL CHALLENGES

Several different ways of categorizing QOS guarantees may be identified. In [21], a distinction is made between deterministic guarantees and statistical guarantees. In the deterministic case, guarantees provide a bound on the performance of all cells (packets) within a session. For example, with real-time traffic, a deterministic guarantee might be that no cells would be delayed more than D time units on an end-to-end basis [23]. With cell loss as a performance metric, the deterministic guarantee might be that no cell loss occurs [12, 9, 10, 5, 6, 11]. Statistical guarantees promise that no more than a specified fraction of cells will see performance below a certain specified value. With real-time communication, a statistical guarantee might promise that no more than $x\%$ of the cells would experience a delay greater than D [13]. With cell loss, a statistical guarantee might be that no more than $x\%$ of the cells in the session are lost [1].

Among statistical guarantees, a distinction is also made between guarantees made in a "steady state" sense, and guarantees which are defined over specific intervals of time. In the former case, if a call were to hold for an infinitely long period of time (an "infinite horizon"), the guarantee would hold with certainty. However, if the session duration is finite, there is some probability that the steady state guarantee would actually be violated. Markovian performance models, such as those in [22, 8, 1], typically fall within the category of statistical guarantees based on an infinite horizon.

When guarantees are defined over intervals of time, the QOS guarantee specifies that a session will not experience a performance level worse than a given value in more than a given fraction of these intervals. For example, an interval-based QOS guarantee might state that "the session will have a packet loss fraction greater than y in no more than $x\%$ of the intervals of time of length I." The relationship between steady-state and interval-based QOS guarantees is discussed at length in [15]. One example considered in [15] is that of multiplexed voice sources over T1-rate lines. It is shown that for the system

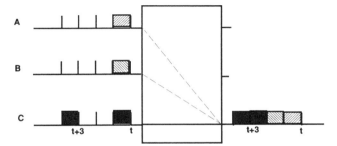

Figure 1. Change in peak rate as a result of multiplexing.

considered, when call holding times are infinite, all calls see a packet loss probability of less than 10^{-3}. However, more than 40% of the calls would experience a packet loss probability exceeding this value if calls were five minutes in length. (Equivalently, more than 40% of five minute intervals in the steady state model have a loss probability exceeding 10^{-3} even though the steady state loss probability is less than 10^{-3}). Interval-based QOS guarantees have also recently been proposed as part of a flow specification in an Internet Request for Comment (RFC) [18].

Characterizing performance (and thus providing QOS guarantees) in BISDN networks poses a considerable challenge for several reasons. First, sources of traffic such as packetized voice and video exhibit correlated, time-varying behavior that is significantly more complex than that of traditional data network sources. Second, since performance (QOS) requirements are defined on an individual, per-session basis, it is no longer sufficient to simply determine the performance of the aggregated network traffic. Instead, performance must be characterized at the finer-grained, per-session basis. Finally, and most importantly, performance must be evaluated in a multi-hop *network* setting. Thus, the complex interactions among sessions must be considered as they interfere with each other as they pass through various network nodes. In this case, one must consider both intra-session and inter-session packet interactions and further consider not only external source inputs to the network but also the session-level departure "processes" at the various queues in order to evaluate performance.

Figure 1 illustrates this last issue. In this figure, three sessions (A, B, and C) arrive to the network at some switch and are multiplexed over a common outgoing link to some downstream switchpoint. In order to evaluate a session's performance at a downstream node, we must clearly be able to characterize that session's departure process from the upstream switch. Note that on input to the switch shown in Figure 1, each session has a peak rate of 1 cell every 3 time slots. Let us assume a work-conserving FCFS multiplexing discipline at the output switch ports (i.e., a cell is transmitted whenever any cells are queued). Let us further assume that among cells arriving in the same time slot, session A cells are transmitted before session B cells, and session B cells before session C cells. As a result of this (not unrealistic) multiplexing discipline, the session C cell arriving at t is output at t + 2, and the session C cell arriving in time t + 3 is immediately output at time t + 3. As a result of multiplexing, session C traffic (which had a peak rate of 1 cell every 3 slots on input) now has a peak rate of 2 cells every 3 slots on output. *The simple act of multiplexing session C with two other sessions has thus resulted in a doubling of session C's peak rate* (defined over a three time-unit interval).

Clearly then, if we are to provide performance guarantees, we must be able either to

. provide multiplexing mechanisms which avoid such changes in traffic characteristics when sessions are multiplexed together, or

. characterize the increases in peak rates, as well as the other changes in a session's traffic characteristics, which occur as a result of multiplexing.

In section 3.1 we outline techniques for providing QOS guarantees using the first of these two altenatives, while the techniques outlined in section 3.2 and 3.3 employ the second alternative.

3. THREE APPROACHES TOWARDS PROVIDING QOS GUARANTEES

3.1 Tightly controlled approaches

In the "tightly-controlled" approaches towards providing QOS guarantees, a non-working conserving multiplexing (queueing) discipline insures that an individual session's output traffic characteristics (i.e., after being multiplexed with other sessions at a switch output port) are the same as that session's input traffic characteristics (i.e., before multiplexing) [12, 8, 9, 10, 6]. An example of a tightly-controlled approach is the so-called "stop-and-go" queueing discipline [9, 10]. This mechanism defines time "frames" and insures that a cell arriving in one "frame" at a switch's input is never transmitted over an output link during the same time frame in which it arrived. Note that in order to satisfy this constraint, a cell may have to be held in the switch's output buffers while the output link is purposefully allowed to go idle. Stop-and-go queueing further requires that a cell *always* be transmitted in next output frame starting after the arriving cell's input frame ends. In order to insure this condition is always possible, arriving sessions may be blocked from entering the network.

With a mechanism such as stop-and-go queueing, a session's traffic characteristics (e.g., its peak rate) are preserved as it passes through the network and consequently performance bounds, such as a deterministic guarantee on the maximum delay experienced by a cell on an end-to-end basis, can be computed in a simple manner. There is, however, a price to be paid for this simplicity of computation. First, a fairly sophisticated, non-work-conserving queueing discipline must be implemented. A second potential disadvantage is that a session admitted to the network essentially "reserves" bandwidth based on *its peak* rate - effectively resulting in a form of circuit switching. As such, classes of traffic with high peak-to-average traffic rates will only utilize the links for a small fraction of their "reserved" amount of time, potentially leaving the links significantly underutilized. As noted in [11], however, it may be possible to utilize reserved, but unused, portions of time to transmit cells from traffic classes which do not require a guaranteed quality of service. Furthermore, if the peak-to-average ratio is not large (i.e., the traffic itself is circuit-like in nature), tightly-controlled approaches towards providing QOS guarantees would be quite attractive.

3.2 Approximate approaches

In approximate approaches towards providing QOS guarantees [22, 8, 1, 7], traffic sources at the network's edge (and within the network) are characterized by relatively "simple" models. An example of such a source model is the on/off source, which alternates between on-periods (during which cells are typically generated periodically) and off-periods (during which no cells are generated) [22, 1, 14]. In order to determine whether or not the multiplexed sources will receive their required QOS, the queueing behavior of the multiplexed traffic streams is then analyzed. In [1, 7] the QOS measure of interest is packet loss; in [8] the measure of interest is maximum delay.

Approximate approaches towards QOS guarantees have both advantages and disadvantages over the tightly-controlled approaches. Perhaps their most important advantage is their simplicity, which makes them well-suited for real-time, on-line implementation. For example, the call admission control mechanism based on the approximate QOS scheme described in [1] can make a QOS computation (to determine whether or not to admit a call, given its QOS requirements and existing QOS guarantees) with a very small number of additions and multiplications. As noted above, tightly controlled approaches require a more complex scheduling discipline; approximate guarantees can be made using simple disciplines such as FCFS. Finally, unlike the tightly controlled approaches described above, approximate approaches are also able to take advantage of statistical multiplexing gains, potentially carrying traffic whose peak rate exceeds link capacity.

There are, however, two important disadvantages. The first is the fact that the approaches are, as their name suggests, "approximate." Empirical evidence in [1] indicates that the QOS computations tend to be conservative, thus providing a guarantee on performance. However, the conservative nature of the approximations has yet to be formally demonstrated, except in the limiting case of small loss probabilities [7]. A second potential drawback is that traffic models, whether at the source or deep within the network, require some form of Markovian assumptions. While traffic at the edge of the network may be reasonably well-approximated by such models [14], it is still unknown whether this is also true for a session's traffic when it is "deep" within the network, having passed through several multiplexers. This remains an important question for future research.

3.3 Bounding approaches

The final approach towards providing QOS guarantees explicitly accounts for the fact that a session's traffic does indeed change each time it passes through a work-conserving multiplexer.

Two approaches toward providing provable performance bounds may be identified - those that provide deterministic guarantees [5, 6, 16, 2, 17] and those which provide statistical guarantees [13, 19, 3]. As noted in section 2, the approaches which provide deterministic guarantees can be used to make statements such as "The delay of every cell from session i is less than x at queue j." The approaches which provide statistical bounds can be used to make statements such as "The probability that a cell from session i has a delay greater than y is guaranteed to be less than z at queue j".

We illustrate the bounding approach by briefly considering the methodology described in [13]. In [13], no assumptions are made about the actual cell interarrival times, as is done in traditional queueing theory. Rather, for each session, a stochastic bound on the number of arrivals in *any* interval of time of length k is specified (typically, for a set of values for k). Given these stochastic bounds on traffic at the edge of the network, bounds can then be computed for each session's traffic after it passes through each multiplexer in the network. Given a characterization of all sources at the "edge" of a given network, and given the routing of sessions, the process of computing performance bounds on a session-level basis is a two-step process. In the first step, all session flows are characterized at each multiplexer; in the second step performance bounds are computed. The two-step procedure is similar in spirit to [5, 6] (although quite different in what is actually computed during each step). In [13], performance bounds on the per-session distribution of delay are computed for a sample 27-session 13-node network, and are shown to be tight for some traffic parameter values but quite loose for others. A number of open research issues, and suggestions for techniques for further tightening the performance bounds, are also discussed in [13].

4. SUMMARY

In this paper we have identified some of the issues involved in providing QOS guarantees and briefly surveyed research in this area. We identified three approaches towards providing QOS guarantees: the tightly controlled approaches, the approximate approaches, and the bounding approaches. The need to provide end-to-end QOS guarantees, while still taking advantage of the resource gains offered by a statistically-multiplexed transport mechanism remains an important, yet largely unsolved problem facing BISDN architects. In a broader sense, although many of the basic hardware technological capabilities for high-speed networks are now becoming available in the laboratory, our understanding of the network's traffic, network control mechanisms, and their performance ramifications is still far behind. On-going research in the area of QOS guarantees represents a significant step in helping to close that gap.

REFERENCES

[1] R. Guerin, H. Ahmadi, M. Naghshineh, "Equivalent Capacity and its Application to Bandwidth Allocation in High-Speed Networks," *IEEE Journal on Selected Areas in Communications,* Vol. 9, No. 7 (Sept. 1991), pp. 968 - 991.

[2] C.S. Chang, "Stability, Queue Length and Delay, Part I: Deterministic Queuing Networks," IBM Research Report RC 17708, IBM TJ Watson Research Center, (Feb. 1992).

[3] C.S. Chang, "Stability, Queue Length and Delay, Part II: Stochastic Queuing Networks," IBM Research Report RC 17709, IBM TJ Watson Research Center, (Feb. 1992).

[4] R. Cruz, "A Calculus for Network Delay and a Note on Topologies of Interconnection Networks," Technical Report UILU-ENG-87-2246, Coordinated Science Lab., Univ. of Illinois, Urbana, IL, July 1987.

[5] R. Cruz, "A Calculus for Network Delay, Part I: Network Elements in Isolation," *IEEE Trans. on Info. Theory,* Vol. 37, No. 1 (Jan. 1991), pp. 114- 131.

[6] R. Cruz, "A Calculus for Network Delay, Part II: Network Analysis," *IEEE Trans. on Info. Theory,* Vol. 37, No. 1 (Jan. 1991), pp. 132-141.

[7] A. Elwalid and D. Mitra, "Effective Bandwidth of General Markovian Traffic Sources and Admission Control of High-Speed Networks," submitted to *IEEE/ACM Transactions on Networking,* 1992.

[8] D. Ferrari and D. Verma, "A Scheme for Real-Time Channel Establishment in Wide-Area Networks," *IEEE Journal on Selected Areas in Comm.,* Vol. 8, No. 3 (April 1990), pp. 368-379

[9] S.J. Golestani, "Congestion-Free Transmission of Real-Time Traffic in Packet Networks," *Proc. IEEE Infocom'90,* (San Francisco, June 1990), pp. 527-536.

[10] S.J. Golestani, "A Stop and Go Queueing Framework for Congestion Management," *Proc. ACM SIGCOMM'90,* (Philadelphia PA, Sept. 1990), pp. 8-18.

[11] J. Golestani, "Congestion-Free Communication in High-Speed Packet Networks," *IEEE Transactions on Communications,* Vol. 39, No. 1 (Dec. 1991), pp. 1802- 1812.

[12] P.M. Gopal and B.K. Kadaba, "Network Delay Considerations for Packetized Voice," *Performance Evaluation,* Vol. 9, No. 3 (June 1989), pp. 167-180.

[13] J.F. Kurose, "On Computing Per-session Performance Bounds in High-Speed Multi-hop Computer Networks," *Proc. 1992 ACM SIGMETRICS IFIP Performance'92 Conf.* (Newport, RI, June 1992), pp. 128-139.

[14] R. Nagarajan, J. Kurose, D. Towsley, "Approximation Techniques for Computing Packet Loss in Finite-Buffered Voice Multiplexers," *IEEE J. on Selected Areas in Comm.,* Vol. 9, No. 4 (April 1991), pp. 368-377.

[15] R. Nagarajan and J. Kurose, "On Defining, Computing, and Guaranteeing Quality-of Service in High-Speed Networks," *Proceedings of INFOCOM'92,* (Annual joint conference of the IEEE Computer and Communications Societies), pp. 2016-2025, May 1992. An extended version of this paper is available as a Technical Report, Dept. of Computer Science, U. of Mass., Amherst MA, 1992.

[16] A. Parekh, R. Gallager, "A Generalized Processor Sharing Approach to Flow Control in Integrated Services Networks - the Single Node Case," *Proceedings of INFOCOM'92,* (Annual joint conference of the IEEE Computer and Communications Societies), May 1992, pp. 915-924.

[17] A. Parekh, R. Gallager, "A Generalized Processor Sharing Approach to Flow Control in Integrated Services Networks - the Multiple Node Case," to appear in *Proceedings of lNFOCOM'93,* (Annual joint conference of the IEEE Computer and Communications Societies).

[18] C. Partridge, "A Proposed Flow Specification", Internet Request for Comments RFC-1363, 1992.

[19] O. Yaron and M. Sidi, "Calculating Performance Bounds in Communication Networks," to appear in *Proceedings of INFOCOM'93,* (Annual joint conference of the IEEE Computer and Communications Societies).

[20] J. Stankovic and K. Ramamritham, *Hard Real-Time Systems,* IEEE Press, 1988.

[21] D. Verma, H. Zhang, and D. Ferrari, "Guaranteeing Delay Jitter Bounds in Packet Switching Networks," *Proc. TriComm'92,* (Chapel Hill, NC, April 1991).

[22] G. Woodruff and R. Kositpaiboon, "Multimedia Traffic Management Principles for Guaranteed ATM Network Performance," *IEEE J. on Selected Areas in Communications,* Vol. 8, No. 3 (April 1990), pp. 437-446.

[23] H. Zhang and S. Keshav, "Comparison of Rate-Based Service Disciplines," *Proc. 1991 ACM Sigcomm Conference,* (Zurich, Sept. 1991), pp. 113 - 211.

HIGH SPEED SWITCHING ALTERNATIVES FOR BROADBAND COMMUNICATIONS

Maurizio Decina

Politecnico di Milano / CEFRIEL

ABSTRACT

At first, service requirements for broadband communications are introduced, including single media (data communications, video communications) and multimedia applications.

The current status of STM (Synchronous Transfer Mode) and ATM (Asynchronous Transfer Mode) high speed switching options in the framework of the B-ISDN (Broadband- ISDN) is briefly reviewed by highlighting both the multirate challenge for STM and the congestion control challenge for ATM.

Then, STM and ATM interconnection networks are classified in the frame of a common taxonomy. Hence, a few types of ATM switching systems are identified and their performance and implementation are briefly discussed.

Finally, a view on today's non-ATM high speed switching alternatives is given, including:
- STM multirate switches,
- variable length datagram switches.

The various switching options are described by taking into account technologies for the implementation of both buffer modules and electronic/photonic space interconnections modules.

HIGHLIGHTS

Asynchronous Transfer Mode (ATM) is the standard adopted by the International Telephone and Telegraph Consultative Committee (CCITT) for implementation of wide-area (see Figure 1) broadband telecommunications networks, namely the Broadband Integrated Services Digital Networks (B-ISDNs).

ATM is based on multiplexing and switching of fixed-size labeled packets of information, called ATM cells. ATM networks operate at data link bit rates in the. order of 150, 600 and 2,400 Mbit/s obtained over fiber transmission facilities.

The attraction of ATM lies into the following main features addressed to single and multi-media broadband applications:

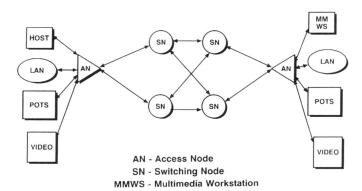

Figure 1. ATM Wide Area Network.

(i) the world-wide offering of a single standard interface;

(ii) the flexibility of accommodating variable bandwidth requests of single and multiple traffic sources: i.e. the multi-rate capability offered by labeled multiplexing and peak-rate allocation;

(iii) the bandwidth saving capability to be obtained by applying statistical multiplexing of cells belonging to bursty traffic sources when mixed with those of stream and variable bit-rate (VBR) traffic sources.

Feature (i) is fully recognized as the golden goal of the standardization efforts and, today, as the trigger for early experiments of ATM interconnection of multi-media workstations in the local areas.

Feature (ii) is addressed as the near end target of ATM wide-area trials. In the present lack of an agreed user-network signaling protocol, networking is envisaged by means of ATM cross connect equipment and the use of Virtual Paths (VPs).

The support of statistical multiplexing [feature (iii)] by ATM wide-area networks to accommodate a wide variety of traffic sources, requires the adoption of suitable traffic management and congestion control mechanisms. Today these mechanisms are still not well understood, in spite of huge research efforts performed world-wide since some years.

The international telecommunications community fully appreciates the complexity of the issue and, to cope with this problem, proposed a large variety of congestion control techniques. The spectrum of congestion control mechanisms for traffic management in ATM networks is shown in Figure 2.

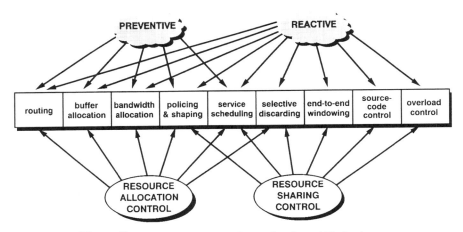

Figure 2. The spectrum of ATM Congestion Control Mechanisms.

Many researchers believe that there is no silver bullet and that control of high-speed packet networks can be obtained only by executing several concurrent mechanisms, including use of priorities in scheduling different services: voice, data (including data services requiring short transport delay), and video (including video services requiring low jitter amplitudes).

Complexity of hardware and software devices, and of operational procedures to control network quality of service to all users (single-media and multi-media) still needs a careful evaluation.

Traffic management and network congestion control represent one of the major challenges in conceiving a pervasive use of ATM to integrate broadband communication services. There are many other issues though that challenge the spreading of ATM technology in future telecommunications networks. The table shown in Figure 3 summarizes these open issues.

A remarkable point is the need for connectionless servers to switch datagrams that are used by most data communication applications.

As far as the previous feature (ii) is concerned, we note that the multi-rate capability could also be offered by exploiting the Synchronous Transfer Mode (STM). The SDH (Synchronous Digital Hierarchy) standard could provide indeed variable bandwidth by suitable aggregations of multiple 64 kbit/s time slots. Although ATM offers a more elegant solution, the STM multi-rate potential was a priori not explored by the standard bodies.

Area	Issue	
Packet Format	Packetizing Delay	Overhead/Payload Ratio
	Header Size and Fields	
Congestion Control	Network "Contracts"	Contract Enforcement
	Priority Schemes	Overload and Flow Control algorithms
	Facility Occupancy	
Inter-Networking and Terminal Interfaces	Echo and Delay	Protocol Conversions
	Terminal Adapters	Interface Costs
Implementation	Large, Low Cost Fabrics	Line Circuit Complexity
Operation Support	Testing	Maintenance
	Operation	Administration
Billing Principles	Billing by Cells or Bandwidth?	
Extensibility	Connectionless Services	Broadcast, Multicast
	Higher Facility Rates	Wireless Applications
	Photonic switching	
User Acceptance	STM and Connectionless Switching or Private Network Alternatives	
	Cost Performance Ubiquity	

Figure 3. ATM Design Challenges.

On the other hand, the use of multi-rate STM in a wide area broadband network is also challenged by some remarkable open issues. For example, the need to conceal the impact of channel bandwidth variations occurring during the connection (that is established by link-by-link routing throughout the network) on user data transmission and transfer procedures. These bandwidth variations are "hitless" in the ATM labeled-cell mode of information transfer.

Finally Figure 4 lists some sample architectures of typical high-speed switching fabrics: three configurations are non-ATM switches, while the fourth is an output queuing ATM switch.

Regardless of the assumed transfer mode (multirate STM/Time Division Switching, Datagram/Variable-length Packet Switching, Space Division Switching and ATM), the switching fabrics are composed by buffer modules and by space interconnection modules. Buffer modules are designed to handle signals formatted by 8-bit bytes (Time Slot Interchange), by variable-length packets (e.g. 10.000-byte datagrams), and by 53-byte ATM cells. In all three implementations the rule of the game is buffer sharing for either signals or commands: buffers are the key elements of high-speed switching fabrics.

Space interconnection modules in broadband switching fabrics are suitable for photonic implementation and they are clocked with timings dictated by the chosen transfer mode and the internal link data rate. Both planar and free-space photonic implementations are conceivable for such modules.

Research is required to optimize the switch architecture for a given transfer mode (there are important alternatives to the samples proposed in Figure 4), to provide multicasting capacity and to guarantee switch overload control.

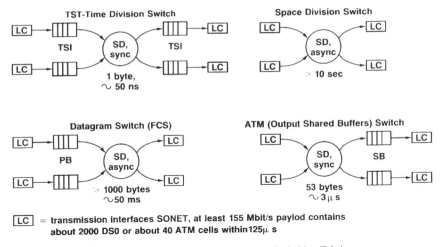

Figure 4. Buffers & Space Switches in Switching Fabrics.

It is not obvious that ATM is superior to multirate STM in terms of, e.g. complexity of buffer and space modules, multicasting capacity and overload control capability.

Moreover, the design of large connectionless switching fabrics (e. g. 1024x1024 ports) needs further attention, since such devices will find application in future broadband networks, regardless of the surrounding network environment (permanent, cross connected, switched, multirate STM or ATM connections).

DESIGN, DIMENSIONING, ROUTING, AND CONTROL OF GIGABIT NETWORKS

Joseph Y. Hui

Rutgers University

ABSTRACT

We re-examine network design, configuration, and routing algorithms for transporting heterogenous services on an integrated broadband network. A layered framework for bandwith management is introduced for path configuration, route planning, dynamic routing, flow control, and cell switching. We then explore the notion of *layered equivalent bandwith* for meeting simultaneously several grade of service (GOS) bounds on the probability of call, burst, and cell blocking. Using this notion of equivalent bandwidth, we consider path dimensioning and configuration for handling heterogenous and time-varying traffic. Also, we shall demonstrate the real-time CAD tool CANeT developed for the design, control, and simulation of multi-service gigabit networks. The tool is an integrated, graphical, and object-oriented software. Interesting dynamical behavior of high-speed networks can be readily observed through animation and color-coded network statistics.

Future broadband services have broad range and ill-defined traffic statistics (figures 2,3). They also differ greatly in grade of service (GOS) requirements. On one hand, the network should facilitate sharing to improve efficiency, while very often, stringent GoS requirements can only be satisfied by bandwidth isolation or dedication. The end-user should be given the flexibility to choose a proper level of sharing. Also, service metering and pricing should be simple and rational.

Future broadband services require traffic control which is layered according to a broad range of time scales (figures 3,4,5). Over periods of months, new transmission and switching facilities are added to meet a long term change in traffic demand. Day to day traffic fluctuations and shorter term changes are often dealt with by configuration paths using cross-connect switches.

For this purpose, the virtual path method is devised for transparent flexible bandwidth management. Shorter term changes, such as call congestion or network failures, could be dealt with by dynamic call routing through preplanned or adaptive route sequencing. Flow control, either by fast reservation or by feedback, maintains a proper arival rate of ATM cells on transmission links.

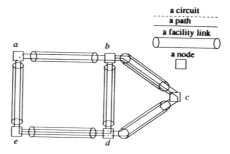

- **Paths simplify routing of calls**

- **A call may use multiple tandem paths**

- **Physical path- Dedicate bandwidth per path**

- **Virtual path- Paths share bandwidth dynamically**

Figure 1. Path Management for Broadband Networks.

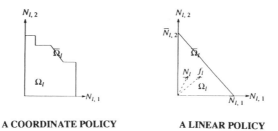

A COORDINATE POLICY   A LINEAR POLICY

- **Call admission policy Ω determines call blocking $P(\overline{\Omega})$**

- **Call layer required bandwidth for given Ω is $\min\limits_{\underline{N} \in \Omega} \hat{C}_{burst}(\underline{N})$**

- **Let S be the set of satisfactory call admission policies with $P(\overline{\Omega}) \leq g_{call}$**

- **Minimize call layer required bandwidth over S.**

Figure 2. Call Layer Required Bandwidth.

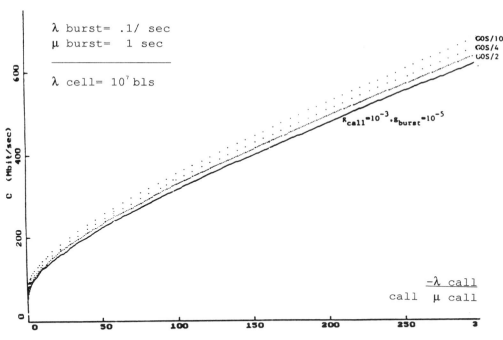

Figure 3. Equivalent bandwith requirments.

Much work remains to be done for engineering the bandwidth requirements, not just for flow control or call admission per se, but to understand how bandwidth management and computation interacts across various layers. For this purpose, a notion of layered equivalent bandwidth is devised [1,2]. At each layer, an admission policy is enforced for the traffic entity at that layer (a cell, burst, call, or path). An equivalent bandwidth is defined based on the GOS requiremnt and traffic statistiscs at the lower layers. Optimal admission policy can be derived for minimizing the worst case bandwidth requirement for the policy. Simple methods for computation are also available (figure 3 and reference [2]).

For path configuration, mathematical optimization problems based on the notion of path layer equivalent bandwidth have been derived for both physical path configuration (figure 4) and virtual path configuration (figure 5). The network design problem is similar to the path configuration algorithms, with multiple traffic periods taken into account. The algorithms are applied to a 10-node network in figure 6, and a comparison of the networking costs for various methods is given in tables 1and 2. It is seen that the virtual path method provides better cost when the paths are constrained to carry only a fraction of the end-to-end traffic, known as the path diversity factor.

Current work on dynamic routing includes an effort to find the cost of routing a call as a function of call bandwidth, holding time, burstiness, link capacity, and link traffic state. Preliminary results show a simple characterization of this function and superb stability and efficiency.

To evaluate overall system performance, an object-oriented and graphical design and simulation package called CANeT has been developed. Using the package, critical links and system dynamics can be readily observed.

Minimize $\quad \sum_e \beta_e u_e$ $\qquad$ (P1)

subject to

$$\sum_{p \in P_\omega} f_p \geq \hat{C}_\omega \qquad \text{for all } \omega \qquad \text{(C1)}$$

$$u_e \stackrel{\Delta}{=} \sum_{p \in Q_e} f_p \leq C_e \qquad \text{for all } e \qquad \text{(C2)}$$

$$f_p \geq 0 \qquad \text{for all } p \qquad \text{(C3)}$$

- β_e is the unit cost for the bandwidth usage u_e
- f_p is the bandwidth assigned for path p
- P_ω is the path group for connecting ω
- Q_e is the set of paths p using link e
- $\hat{C}$ is the equivalent call layer bandwidth
  ; the capacity of link e

Figure 4. Configuring physical paths.

Minimize $\quad \sum_e \beta_e u_e$ $\qquad$ (P3)

subject to

$$\sum_{p \in P_\omega} r_p = 1 \qquad \text{for all } \omega \qquad \text{(D1)}$$

$$\hat{C}_e (\mathbf{g} , \sum_{p \in Q_e} r_p \mathbf{h}_{\omega(p)}) \leq u_e \leq C_e \qquad \text{for all } e \qquad \text{(D2)}$$

$$r_p \geq 0 \qquad \text{for all } p \qquad \text{(D3)}$$

- β_e is the unit cost for the bandwidth usage u_e
- r_p is load sharing probability for path p
- P_ω is the path group for connecting ω
- Q_e is the set of paths p using link e
- $\hat{C}_e$ is the equivalent call layer bandwidth for calls using e

Figure 5. Configuring virtual paths.

Table 1 (One service)

| ω | Configuration Cost ($) | | |
	VPC	PPC	(%)
1.0	9872	15458	56.6
0.8	10722	16946	58.0
0.7	11121	17686	59.0
0.6	11507	18591	61.6
0.5	11881	XX	

ω : Path Diversity Factor
(%): Cost saving by using VPC
XX : There is no Feasible Solution

Table 2 (Two service)

| ω | Configuration Cost ($) | | |
	VPC	PPC	(%)
1.0	9183	14051	53.0
0.8	9982	15390	54.2
0.7	10358	16079	55.2
0.6	10721	16792	56.6
0.5	11078	XX	

REFERENCES

[1] J. Hui, M. Gursoy, N. Moayeri, R. Yates, "A layered broadband switching architecture with physical or virtual path configurations," IEEE J. on Selected Areas on Comm., Vol. 9, Dec. 1991.
[2] J. Hui, "Layered required bandwidth for heterogeneous traffic," Proceeding of IEEE Infocom, Florence, Italy, May 1992.

APPLICATIONS FOR BROAD BAND COMMUNICATIONS

Stephen B. Weinstein

Bellcore

ARE THERE ANY?

Are there any applications for broad band communication, at rates of tens of megabits per second and higher? There are good reasons to question the assumption that a large number of people will want anything more than those activities - let's call them "applications" - that can run very well on existing lower rate communication networks or on existing broadcast systems. The concept of broad band communication implies a high degree of interactivity at high transmission rates, although it does not exclude broadcast and lower-rate interactive applications.

In particular, will residential users of communications services ever want anything other than telephone and CATV-like video entertainment services? Telephone networks support interactive data communication, using modems, in addition to a wide range of voice-oriented applications. Many CATV systems have increased capacity and offer a wide range of programming. It is not clear that consumers are willing to spend more on video entertainment.

Advances in video and digital communication technologies add weight to the argument that existing communication facilities may have an extended life. Compressive video coding makes possible the transmission of still and moving images at rates an order of magnitude lower than many thought possible a few years ago at the quality levels that are being achieved. Video telephony, albeit with crude moving video, is being marketed at a rate of 9.6 Kbps (not including the audio). Codecs producing interactive video of decent quality are becoming available for the 64 Kbps ISDN Basic rate interface. At rates of 384 Kbps to 1.5 Mbps, very impressive interactive video (e.g. with the TSS (formerly CCITT) H.261 recommendation) and distribution (one-way broadcast) video (e.g. with the MPEG standard) are now possible. Digital HDTV appears to be on the way to becoming a standard at transmission rates of the order of 20 Mbps, fitting into existing broadcast and CATV channels. Color still images of quite good quality can be transmitted in a very few seconds at the 64 Kbps ISDN rate.

Worldwide Advances in Communication Networks
Edited by B. Jabbari, Plenum Press, New York, 1994

Advances in digital communication parallel and complement the advances in video coding. Applying digital signaling techniques that originated in voice channel data communication, ordinary copper subscriber telephone lines can be used at surprisingly high rates. There are many possible combinations of copper and fiber optic facilities. The Asymmetric Digital Subscriber Line (ADSL) is typically described as use of a single subscriber line for data transmission at 1.5 Mbps from telephone office to subscriber in addition to normal two-way voice communication (POTS). Bringing fiber a little closer to the subscriber, a single line might carry POTS, two-way 384 Kbps, and one-way 1.5 Mbps, or POTS and one-way 3-4 Mbps, or even 6 Mbps as some have suggested. Two-way transmission, as contemplated for the High-Speed Digital Subscriber Line (HDSL), could carry 1.5 Mbps in two copper telephone lines, or higher with a tandem combination of copper and fiber lines. In CATV networks, digital distribution systems are being developed that support signaling at about 24 Mbps in each 6 MHz transmission channel, sufficient for six 4 Mbps coded video signals in a channel that today is used for only one.

The questions go beyond the capacity of existing networks. One can also ask if physical media - videocassettes, videodisks, CD-ROMs - won't handle most bulk information transfer needs. And isn't most computing done locally, needing only local communication? Won't progress in powerful, people-oriented personal computers and workstations reduce the need for intelligent, broad band communication networks to access such capabilities remotely? Finally, in considering a broad band public network that can do much more than today's telephone network, one must ask if quality of service for a wide range of media and service types can be reconciled with network efficiency and acceptable pricing.

THE NEED FOR BROAD BAND COMMUNICATION

So why does broad band communication have a future? Essentially because the transition to high-volume, more visually-oriented traffic, and the desire and need to reach out to widely dispersed people and information resources, will not be satisfied by existing communications facilities. Let me make some predictions.

First, "telepresence" through multimedia interpersonal communication is likely to capture the public imagination and generate new markets for broad band communication, including special network-mediated services such as audio and visual composition. It will also encourage mass production of user equipment that embody advanced computing, media, and communications technologies. Large-screen teleconferencing, desktop teleconferencing, and multimedia messaging are three promising formats.

Large-screen teleconferencing is the use of a life-sized display, typically a wall-mounted, wide-aspect screen at least three feet high and five feet wide, and open, hands-free, high-quality audio, to impart a feeling of naturalness to a conversation between small groups at the participating locations. Classrooms of the future could create an educational experience in which students have meaningful contact with instructors, subject experts, and other students in other places. The wall in a future hotel room might (almost) bring a traveler home to family and business colleagues. Photographs, documents, and computer applications could be shared. Desktop (or perhaps kitchen counter top) teleconferencing implies audio and visual connections from a personal computer, digital assistant, or workstation to other people. The audio/visual "windows" would work in parallel with other applications, possibly shared, on the personal device. Information resources found locally or from distant sources could be brought into, or derived from, the interpersonal teleconference.

Multimedia messaging is the use of more than one medium for deferred communication. The media possibilities include text (as in electronic mail), voice (as in today's voice mail), document bit maps (as in facsimile), data files, color photographic images, audio, moving video, executable computer code, and whatever other media may be available and useful to the sender. Messaging is, as we all know, very useful for communication with people who are hard to reach directly. But this modern convenience is only one motivation. Messaging has a long history in written correspondence and a secure place in human psychology. A tourist of the not very distant future might carry a video camcorder with a built-in personal communications network (PCN) radio transmitter, and appropriate application software. Taking a five or ten second video clip, the tourist will attach an audio comment (wish you were here?) and send this "video postcard" off to someone back home. The address might be read from a directory previously stored in the camcorder or a pocket digital assistant, or handwritten with an electronic stylus and recognized by software resident either in the tourist's equipment or in a server provided by the electronic messaging service. The video postcard could be delivered directly to recipients with ADSL, advanced CATV, or another personalized video delivery capability, or indirectly via a transcription to videocassette to others. One video postcard may not call for broad band communication facilities, but millions of video postcards and longer "letters" would.

In addition to new kinds of interpersonal communication, people will want the greater choice in content, quality, and control of visual and multimedia information facilitated by intelligent (software-controlled), broad band communications. This will both complement and compete with physical media such as CD-ROMs and videodisks. It will never be possible for physical media to provide the breadth of materials, currency of information, processing and composition capabilities, and immediate delivery that are available through broad band information networking. Does the corner video store have the not-so-popular film you want to see, or the local library the extensive reference materials (including video where appropriate) that you need for a paper? Can you obtain a digital video clip in either MPEG or HDTV format? I don't know about video stores, but public libraries are already making progress toward transforming themselves into information gateways and guides rather than information repositories.

The fact is that digital video processing technologies are opening near-term markets for video applications on lower-rate access facilities, and long-term opportunities for multimedia telepresence and information applications using information-rich signals. Future digital television sets will have the processing "intelligence" for local management of sophisticated multimedia information applications, and the future public network will provide intelligent communications platforms for such applications. An educational application, for example, would offer students in classrooms and at home access to programmed learning modules, reference materials from different sources, and collaboration and composition tools, in addition to direct contact with instructors and other students.

Making intelligent media and computing equipment inter work with each other and with an intelligent communications network such that everyday applications are easy to deploy, easy to use, secure, and inexpensive, while remaining profitable and effective for the communications and applications providers, is a major challenge for our industry. A customized multimedia information retrieval applications, with personalized filtering of information from a number of subscription sources, illustrates technical/economic tradeoffs that might not be evident from a superficial consideration. Should information filtering be provided near the user for privacy and quick response, near the provider for protection of intellectual property, or somewhere in between for minimization of traffic volume? Should subscriber-network signaling include identification of traffic type, to facilitate special services such as format conversions or screen composition, or only more abstract specification of quality of service? We are truly entering an era of distributed systems, with elements provided by diverse and sometimes competing entities, where performance optimization will require new definitions of open systems and of the nature and functions of standard interfaces.

APPLICATION EXAMPLES

The broad band communications requirements can be described for a few specific application examples. For point-to-point large-screen group teleconferencing, 384 Kbps transmission will probably provide the minimum acceptable audio and visual quality. A research prototype created at Bellcore provides the HDTV aspect ratio (16:9) but uses conventional NTSC signals and special processing circuitry. In the future, users may wish to upgrade to digital HDTV, at a transmission rate of about 20 Mbps, for a more pleasing large display and for easy mixing of good quality graphics and text with views of other people. For multipoint conferencing, special multipoint control units, including audio and video composition features that allow users to customize their views of other locations and participants, and allowing combination of different types of signals from different locations, may be required for success of the application. In order to allow a greater choice of user options in selecting views of groups and individuals at other locations, panning, and zooming in, the signals provided from each location may be considerably more information-rich (and hence demanding of bandwidth) than the capacity required for a particular view. Broad band communication is useful not only for higher resolution, but also for greater choice and personalization.

Medical image exchange is expected to provide a substantial quantity of traffic on broad band networks. Data shown by H. Blume of Philips Medical Systems suggest an annual production volume, in a typical 500-bed hospital, of almost 2,000 Gbytes of medical image data. Hospitals will be cautious about use of compressive coding because of the liability possibilities. The minimum daily volume of image transactions is estimated at about 40 Gbytes, corresponding to an average 6.5 retrievals of a daily production of about 6.75 Gbytes. The collaborative examination and manipulation of medical images, by (for example) specialists in consultation, will require applications for this purpose that make efficient use of communication resources while receiving a high quality of service. Half-second retrieval times, security, and quick and easy setup and tear down of multiparty sessions will be demanded. Broad band communication provides both the capacity needed for large quantities of image transfers and the low latency for rapid retrieval, where most of the delay may be due to the storage system rather than the information network.

Video on demand is a special case of customized information retrieval that illustrates the advantages high-speed Asynchronous Transfer Mode (ATM) and a network-based applications platform could have. By high-speed burst downloading of short digital video segments to users, at roughly 100 times the normal playing rate, multiple users can be statistically multiplexed while each user has the illusion of an exclusive connection with full control of viewing comparable to control of a videocassette in a VCR. In a research prototype configured at Bellcore, the user can stop on freeze frame, fast forward scan, rewind any amount, or jump to anyplace in the video program. The application platform, at the serving public network office of the subscriber, can be programmed by the service provider to realize any number of service, advertising, and billing options.

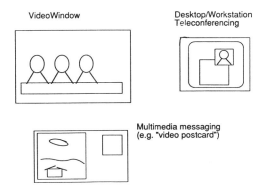

Figure 1. Some predictions.
"Telepresence" through multimedia interpersonal communication will capture the public imagination and eventually a substantial market.

CONCLUSIONS

The conjunction of computing, media, and communications technologies is leading to profound changes in the ways people obtain and use information, work together, and maintain daily contact with one another. The challenge for communications providers and manufacturers of communicating equipment is to provide a common, well-understood distributed environment supporting a wide range of potential applications, easy for applications creators and providers to use. The public network must consider overlay vs. integrated architecture's, provision of both complex capabilities (e.g. applications platforms) for sophisticated users and mass-market "teleservices" for a large subscriber population, and an evolution of services from those offered on present facilities to those on a broad band, intelligent network that build new markets without losing existing ones.

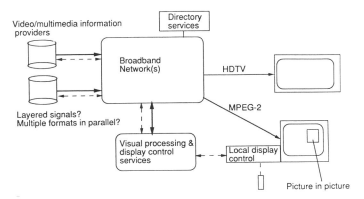

Figure 2. Some predictions.
People will want and use the greater choice in content, quality, and control of visual and multimedia materials facilitated by intelligent, broadband communications. This will both complement and compete with physical media.

43

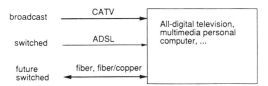

Figure 3. Some predictions.

Digital video processing technologies are opening near-term markets for video applications on lower-rate access facilities, and long-term opportunities for multimedia telepresence and information applications using information-rich signals.

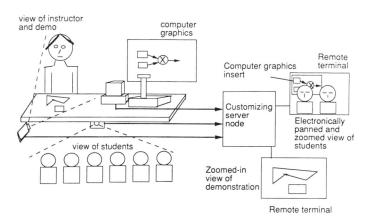

Figure 4. Distance Learning: Individualized Uses Of Information-Rich Signals.

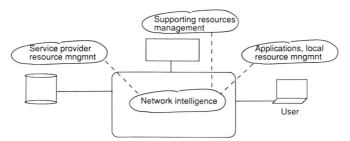

Figure 5. Some predictions .
Computing facilitating information management, visualization, and distributed applications will itself be distributed. "Intelligence" in end equipments and communication networks will interoperate for optimal cost/performance, and rarely be all "central" or all "at the network periphery".

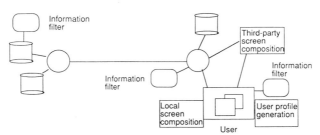

• Alternative information filtering locations:

   - Near user for privacy and quick response
   - Near provider for protection of intellectual property
   - Somewhere inbetween or distributed for traffic minimization

Ref: H. Bussey & S. Weinstein, "Communication session management for distributed multimedia applications", Proc. Int'l Zurich Seminar on Digital Communication, March, 1992.

Figure 6. Example: Customized multimedia information retrieval.

Figure 7

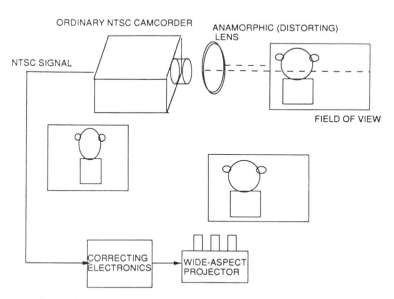

Figure 8. Video processing for HDTV Aspect Ratio Using NTSC Signal.

[From DEC Product Insight, Sept/Oct 1991
courtesy Digital Equipment Corporation.]

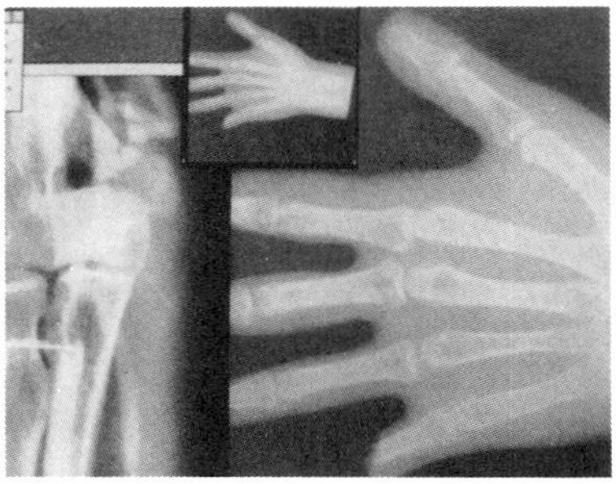

Figure 9. Radiological Imaging.

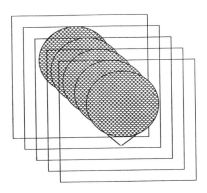

8 Mbytes/image x 2 images/sec = 128 Mbps

Figure 10. Fast Scanning of Radiological Images.

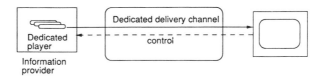

WHAT IS IT?

Ideally, telecommunications access to any available form of a video program or segment, when you want it, under your full individual control.

Costs high for dedicated players and high network throughput.

Figure 11. Video on Demand.

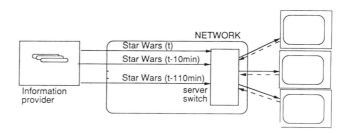

- Program re-broadcast at T intervals, nT=total running time, on parallel channels

- Viewer can skip forward or backward in T quanta

- Unlimited number of viewers can receive any time-shifted version

- Number of read heads/single program is n

Figure 12. Near-Video on Demand For Program With Many Viewers: Pay-Per-View on Frequently Repeated Schedule.

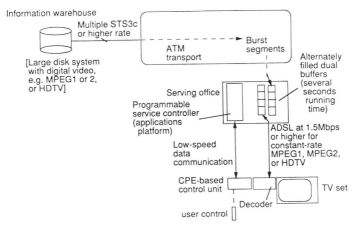

Figure 13. Video on Demand Delivered by ATM.

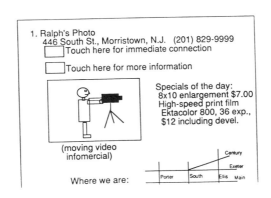

Query: "A few nearby photo finishing stores"

Figure 14. Concept for Electronic Yellow Pages.

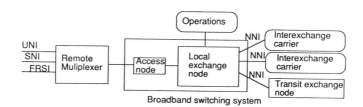

UNI - B-ISDN user-network interface
SNI - SMDS subscriber network interface
FRSI - Frame relay service interface
NNI - Network-network interface

Ref: S.M. Walters, "A new direction for
broadband ISDN", IEEE Com.Mag.,
September, 1991.

Figure 15. Access to Broadband Communication: B-ISDN Architecture.

MULTIMEDIA APPLICATIONS AND THEIR COMMUNICATIONS NEEDS

Stewart D. Personick

Bell Communications Research, Inc.

It is a pleasure to be here to speak at this symposium amongst so many old friends and new friends who are helping to shape the technological underpinnings of the emerging information age.

As I listened to the earlier speakers, I noted that we are all becoming a bit more conservative in what we say with the passage of time. I think there are a number of reasons for this that are worth noting.

We are all getting older, and I guess this has a way of knocking off some rough edges, in and of itself. Some of the dreams we have all articulated in the past have not come true exactly as we imagined, and this has a somewhat sobering effect. Some of the dreams we have all articulated in the past are coming true, and we are getting closer to the money, which makes us more cautious about what we say. As these dreams become reality, everything becomes more complex, and it is difficult to make simple, blunt statements without a myriad of caveats; thus we tend to speak in more abstract terms. Finally, having gotten older and established ourselves, people actually take what we say a lot more seriously; and this gives us all a good reason to be cautious.

Having said all of that, I will make some specific remarks on what I perceive to be an important subject, and one where there is a great deal of focus in my research organization in Bellcore. This has to do with understanding what people will really be doing with computers and communications in the future, and how their needs for communications capabilities will influence the telecommunications networks that they select to support their applications.

Let me clarify a bit what I mean by a need for a communications capability. I'll do this in a very familiar context. One of my communications applications is to be able to carry on a voice conversation with someone who is not in the same room with me. What are the communications needs that go along with this application? Well, I would like to have some way of notifying the person with whom I wish to speak that, indeed, I wish to do so, wherever in the world that person happens to be. I would like that person to be able to receive the notification, and to be able to know that it is I who is calling. I would like that person to be able to conveniently decide whether he or she wishes to speak with me, and to have the ability to conveniently reply to my request, one way or the other, in a manner which lets me know his or her decision. If he or she does want to speak with me, then I would like to be able to conveniently begin a high quality voice conversation. There are other things I could list, but I'll stop at this point.

As you can see, my needs in the context of having a conversation with someone who is not in the same room with me are only partly met with today's telecommunications networking facilities. Typically I make a call to a particular telephone rather than to a person, hoping that the person will be near that telephone. Unless I utilize a paging service, or an emerging capability like calling name delivery or user-to-user signaling in ISDN, I can't provide much information to the person I am calling to identify myself or the purpose of my call. Without user-to-user signaling, the person receiving the call can't provide any response (e.g., I'll call you back in a few minutes) without actually picking up the phone and starting a voice conversation. Also the audio quality of the conversation is limited to the capabilities of typical voice grade telephony.

Thus the networking capabilities available today, although continuously improving, fall short of meeting all of my communications needs in this application. From a technology perspective, I'm sure that we can all think of ways of implementing new and improved networking capabilities to fill in the gaps between telephony as it exists today, and what I described as my communications needs. However, filling in those gaps involves very large expenditures, and thus an important issue is the determination of which of my stated communications needs are most important in really helping carry out my desired application. In an increasingly competitive environment, the difference between winning and losing in the marketplace will be heavily influenced by a network provider's ability to understand what people will really find important in their applications, and in creating the most effective and efficient means for serving those needs.

This brings me to the topic of multimedia applications and their communications needs. Unlike the familiar application I described above, which has been used by billions of people for over a hundred years, multimedia applications are in their infancy. Around the world, research laboratories have implemented prototypes of multimedia applications for information access and interpersonal communication. The concept of distributed cooperative work amongst members of geographically distributed teams, or between businesses and their customers, or in the context of distributed leisure activities (e.g., the "work" of children is "play") has led to numerous implementations of interconnected multimedia workstations with auxiliary shared electronic whiteboards and other mechanisms to allow people to do at a distance what they normally do together in the same room.

One of the most important aspects of this research is to understand in an objective way what is really required to create the feeling of presence in distributed cooperative work applications of various kinds, and what the most important communications needs of those applications will be. These insights are essential if the investors in the future telecommunications networking capabilities in the U.S. and around the world are to make those investments wisely--in a way that most effectively serves the most important needs of emerging applications in a timely and economic way.

One of the classes of applications of computing and communications we have been studying in some detail is multimedia desktop teleconferencing as a form of computer and telecommunication supported cooperative work. We have placed multimedia workstations, cameras, electronic whiteboards, and shared computational objects (e.g., shared customized views and access to a common graphical object) in the offices of our researchers, to understand how these capabilities can allow them to do such things as "drop in" on each other for spontaneous interactions, very much as people can drop in on each other by walking down a hallway and joining people in their offices if the door is open. We have studied the privacy aspects of these types of applications, as well as the relative importance of various aspects of the total application in creating the needed "presence" to conduct various types of work. Not surprisingly, it is relatively easy to create an electronic environment that allows people to do such things as plan for a meeting or ask a simple question. What is harder is to provide the richness of simulated physical presence needed to discuss abstract concepts and to jointly view and edit complex documents or drawings.

We are making progress, and there are some preliminary results that may be surprising, even controversial. For example, the ability to share ad-hoc drawings and to jointly view and edit text and graphical objects, as well as the need to pay proper attention to privacy and ease of use, appear to be more important than the ability to provide artifact-free full motion video images of the participants. This is not to say that video is unimportant or unappreciated; but rather other aspects of these applications may have a bigger impact on their usefulness and their degree of actual use.

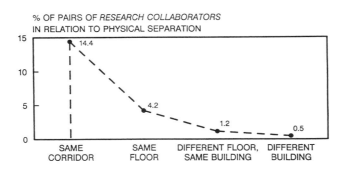

% OF PAIRS OF *RESEARCH COLLABORATORS* IN RELATION TO PHYSICAL SEPARATION

Robert Kraut, Jolene Galegher, & Carmen Egido, June, 1988

Figure 1

In our desktop teleconferencing application, people have found it useful to connect to information sources--seminars or TV, for example--using the same techniques they use to contact other people. To do so requires what might be described as a real time directory that provides users with appropriate permissions access to a list of communications sessions actually in progress and which are open to others to join. Thus a user can discover an existing multimedia conversation, and proactively join it. The apparent need for such capabilities may have a significant impact on how future multimedia-capable networking facilities will function.

In summary, as we move closer to the realization of our vision of information age computing and communication capabilities to enrich peoples lives, to provide access to quality education and health care, and to increase the effectiveness and efficiency of businesses in meeting the needs of their customers and in collaborating with their suppliers, distributors, and strategic partners...we must place increasing emphasis on understanding what is important in the multimedia, multipoint, personalized applications which are implied by this vision, and what networking capabilities are needed to support the communications needs of these applications.

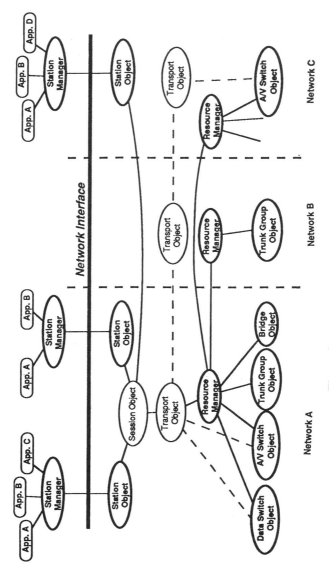

Figure 2. Network Software Architecture.

REFERENCES

1. Fish, R. S., Kraut, R. E., Root, R. W., Rice, R., "Evaluating video as a technology for informal communication", CHI 1992 Conference Proceedings (Monterey, CA, May 3-7, 1992), P. Bauersfeld, J. Bennett, & G. Lynch (Eds.), ACM, New York.
2. Cool, C., Fish, R. S., Kraut, R. E., Lowery, C. M., "Iterative design of video communications systems", CSCW 1992 Conference Proceedings (Toronto, Canada, Nov. 3-7, 1992), R. Kraut & J. Turner (Eds.), ACM, New York.

APPLIED MICROCELL TECHNOLOGY IN THE PCS ENVIRONMENT

W. C. Y. Lee

PacTel Corporation

ABSTRACT

A new scheme to be applied in the PCS environment in order to increase system efficiency as well as reduce the size of portable units and prolong talking time.

INTRODUCTION

As life styles change in the 90's, people are continuously on the move as illustrated in Figure 1. A wireline system is not viable for their communication needs. Hence, wireless means of communication are more suitable. Everyday the average person may be in an office, meeting room, library, shopping mall, supermarket, gymnasium, etc. Thus, conventional concepts of distinguishing market segments cannot apply. The PCS (Personal Communication Service) environment can be described as a wireless and personal communications environment. PCS must serve large cells for mobile communications and small cells for portable-unit communications; outside-building and in-building; above ground and underground; in the air and over water; fast motion, slow motion or standstill; or may even use communications via satellites. The difficulties associated with PCS communications are excessive path loss; multipath fadings due to moving terminals; and time delay spread of multipath wave arrival due to human-made structures. Thus, providing a viable personal communication system becomes a challenge.

DEFINITIONS OF PCS

PCS has never been clearly defined up to the present. Here is one definition which can be viewed by either system operators or subscribers:

From the System Operator's View

a. Coverage - The system operators would like to provide coverage wherever PCS subscribers are present.

b. Service - Service quality should be good. Calls should have low blocking probability and low "drop call" rates. Services such as voice, data, FAX, user's locations, etc., should be offered.

From the Subscriber's View

A personal unit means a unit which a person can carry with them at all times. Therefore, it should be light weight, small in size, easy to carry and should provide long talk time along with many of the features listed above. In this case, power consumption of a unit should be minimized so that the battery can last longer. One arrangement is to have the power emission close to the PCS unit as shown in Figure 2. As a result, the power consumption is lowered and longer talk time can be achieved. Thus, we do not need to depend on the battery industry to deliver a more powerful battery for consumers.

Subscriber's Interest

Subscribers would like to carry one PCS unit and use it wherever they may be. They do not like to carry many different units such as cellular, PCN, CT2, FAX, fleetcall, pager, etc., and call themselves PCS users.

Figure 1. Mobility.

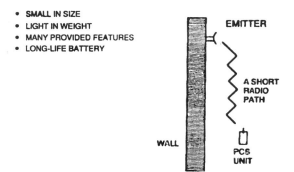

Figure 2. PCS Subscriber Unit.

PRESENT PORTABLE-UNIT SYSTEMS

Today's cellular system is not quite a PCS system. It utilizes cell sites to connect calls to portable units inside buildings. This connection causes two problems. First, the signal has to penetrate through the building walls. The average penetration loss through buildings in the Chicago area is around 18 dB, in Los Angeles 22 dB, and in Tokyo 28 dB. These losses vary in different areas but they remain quite high. These differences are due to earthquake situations. In Tokyo, the high-rise buildings have been constructed with dense mesh-type steel which provides protection against earthquakes. Thus, the signal penetration loss is very high. The other problem is that portable unit coverage is much smaller when compared with mobile unit coverage. The power of portable units is 0.6 watts as compared with the power of mobile units which is 4 watts. It is difficult to carry portable unit calls in the same manner as mobile unit calls. However, portable units can go up to higher floors. The higher the floor, the further the coverage is for portable units as shown in Figure 3. Sometimes portable unit calls on high floors will interfere with ground mobile units. Due to this phenomenon, it is difficult for a cellular system to serve both portable and mobile units and still provide adequate coverage with good performance in congested areas.

NEW CONCEPT OF PCS AND IN-BUILDING COMMUNICATIONS SYSTEMS

Signal penetration into buildings in today's cellular systems from outside cells is not a good approach, because it requires more power and generates more interference. Moreover, when a frequency channel is transmitted from an outside cell into the building, the signal will be received throughout all the floors, which hinders the transmission of two messages on the two identical frequency channels to two different floors. Thus we should consider a different approach in designing an in-building communication system. We can treat the building walls as a natural shelter for preventing the signal from penetrating in or out of the building. The signal will then be delivered into the building by optical cable, T1 cable or microwave links as shown in Figure 4. In each floor, the same set of frequencies will be used. This means that the same 20 cellular frequency channels are used in each floor (see Figure 5). Since the isolation between floors is about 20 dB, we can use the same set of frequencies on different floors without causing cochannel interference. In this case, 400 voice channels can be provided for a 20-story building. Moreover, the same set of frequency channels can be used in the next building. The isolation from two building walls helps prevent interference. To further reduce the interference we can use the new microcell architecture (see Reference 1) in each floor. The philosophy of microcell allows the power to follow the mobile or portable units. Each floor will have three or four zones as shown in Figure 6. Each zone site will cover a small area on each floor with a zone scanner located at the base (which may be located outside). The system intelligently knows where the portable unit is located. The assigned zone site only needs to transmit at a minimum power level to the portable unit, thus reducing the power level on the whole floor. Lowering power means lowering interference, which in turn allows the same set of frequency channels to be used in an adjacent building with no cochannel interference.

In this new system, each building is treated as one cell site. Therefore, one cellular portable unit can be used for inside as well as for outside buildings without modifying the unit. Each floor is treated as a sector, and each building can have as many sectors as the number of floors. All the sectors use the same set of frequencies, which is a spectrum efficient system.

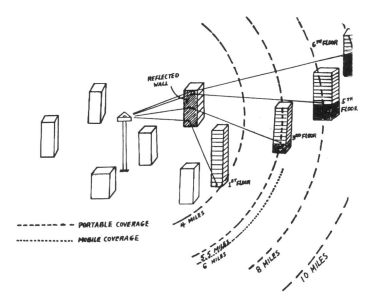

Figure 3. PORTABLE AND MOBILE COVERAGE.

SIGNAL DELIVERY SYSTEM

In the new PCS system, it is important to provide a practical signal delivery system. A signal delivery system is developed by situating all the radios in one large storage location as shown in Figure 6. Those radios, which will serve a designated building, will be maintained by one cell-site controller. The cellular signals that will serve one floor from a set of radios (roughly spread over a spectrum of 12.5 MHz) will be upconverted to either microwave frequency or optical frequency at the base, then downconverted to the cellular frequencies when reaching that particular floor of the building. These boxes are called translators. The details of the signal delivery system are described in [1] and [2]. All the radios are located at the base. With the delivery system, the base can reside

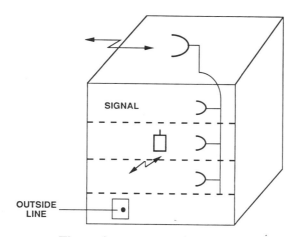

Figure 4. Signal Within the Building.

anywhere and the buildings do not need to provide space for radios. The network for using fiber cables or microwave links can be easily modified or moved around, which is important for PCS.

ATTRIBUTES OF THE NEW SYSTEM

Below are some of the attributes of the new system:

1. High spectrum efficiency;

2. Minimal change in cellular portable units;

3. No modifications in the MTSO;

4. Low cost and simple changes at the cell sites;

5. Smaller size and longer talk time for portable units;

6. Portable units can be plugged in to power amplifiers in vehicles for larger cell coverage.

Therefore, we can deduce from the aforementioned that this is a suitable network for cellular PCS.

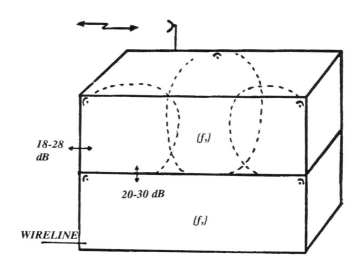

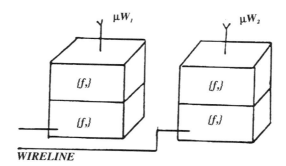

Figure 5. Concept of inbuilding Communications.

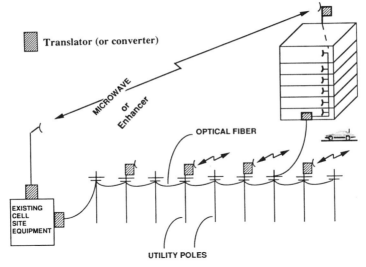

Translator (or converter)

- Upconverting 900 MHz GSM signal to optical signal
  (or microwave signal) and down converting optical signal
  (or microwave signal) to 900 MHz

Figure 6. PACTEL MICROCELL DELIVERY SYSTEM.

REFERENCE

[1] W. C. Y. Lee, "An Innovative Microcell System," Cellular Business, December 1991, pp. 42-46.

[2] W. C. Y. LEE, Mobile Communications Design Fundamentals, John Wiley, New York, 1993, p. 337.

ISSUES IN MICROCELLULAR COMMUNICATIONS - CDMA VERSUS TDMA

R.L. Pickholtz and B.R. Vojcic

Department of Electrical Engineering and Computer Science
The George Washington University
Washington, DC 20052

ABSTRACT

In this paper we briefly outline some current issues in personal communications networks and then compare the capacity of CDMA and TDMA in a microcellular radio channel. We calculate the performance of CDMA and TDMA in a frequency selective Rician channel, which appears to be representative of the microcellular environment. The effects of cell sectorization and forward error correction are analyzed. Achievable capacity improvements due to the voice activity, in both schemes, are accounted for. In the case of TDMA, both equalized and unequalized performance are considered.

INTRODUCTION

Personal communications networks (PCN) are considered as the means to communicate with *anyone, at any time, anywhere* .... In other words, to achieve the ultimate goal of universal communication services. The personal communications services (PCS) have to include the transmission of voice, data and images using a universal personal communications terminal. The universality objective implicitly suggests a common technology for the home, business and public access sectors. These goals, with diverse channel conditions, user densities and traffic requirements pose a challenge to achieving the universality for which a market need is evident. The short term business needs favor different technologies for the three different access sectors. That is, in order to achieve competitive service and terminal prices at this moment, the industry is making every effort to fulfill the short-term goals and penetrate the market. However, developments in digital signal processing (DSP) technology will provide the means for the so-called multimode software radio, which can satisfy all the requirements at a competitive cost. Still, there is a whole spectrum of unresolved issues regarding PCN.

One of the main requirements for the development of PCS is a coordinated radio spectrum allocation on a world-wide basis. A major breakthrough in this area was made at the 1992 World Administrative Radio Conference (WARC'92). The 1700 to 2690 MHz frequency band was exclusively allocated for terrestrial and satellite mobile services, that is, for PCS. However, the use of other frequency bands for PCS is not precluded. This opens new opportunities for the development of PCS.

In addition to the radio spectrum and technology availability in the near future, there is a need for standardization in the area of PCN. The priorities in standardization, on a world-wide basis, are seen in signaling and numbering plans, roaming, privacy, authentication and a common air interface. With respect to this matter, there is a disagreement whether the standardization process should be regulated by the International Telecommunication Union (ITU) or market driven. The later option seems to be the prevailing opinion in the USA, but, both approaches have to be implemented in a balanced way.

One of the main technical issues, still unresolved, is the channel access. A basic question is which channel access method would provide the largest radio capacity and, hence, maximize the number of potential users that can be served, with the limited natural resource, the radio frequency spectrum. A possible radio capacity improvement would have a direct impact on the service and technology development per user cost. Technically speaking, the radio capacity of a multiple access scheme is dependent upon source and channel coding, radio channel characteristics, antennas and a number of related issues. From a strictly theoretical point of view, the three multiple access techniques, frequency-division multiple-access (FDMA), time-division multiple-access (TDMA) and code-division multiple-access (CDMA) are equivalent with respect to the radio capacity. That is, the number of orthogonal users depends on the dimensionality of the signal space, the time-bandwidth product [1]. In a real world, their capacity may differ from each other due to a number of practical reasons. Indeed, the debate regarding the choice of multiple-access techniques for different terrestrial and satellite communication applications started in the early 80's and is getting a new momentum now. Although different forms of narrow-band and wide-band TDMA were accepted as next generation standards in the USA, Europe and Japan, CDMA is gaining new advocates every day and has been recently accepted as the second standard in the USA.

There are numerous examples of the performance comparison of TDMA and CDMA. Without any intention of discussing all of them, we will briefly outline a historical overview of this controversy. It was shown in [2], for cellular radio assuming a non-selective fading channel, that CDMA is less spectrally efficient than either FDMA or TDMA. The same conclusion for non-fading satellite communications was drawn up in [3]. These and similar results (see [4]) have certainly contributed to a widespread opinion that CDMA has no potential applications in the non-military area. However, the opposite opinion has also existed. Cooper and Nettleton showed in [5] that the efficiency of a spread spectrum scheme may exceed the efficiency of classical narrow-band schemes by a factor of five. More recently, with the commercial availability of the spread spectrum technology, some new opinions in favor of CDMA have emerged. The analysis in [6], for mobile satellite communications, indicates that the capacity of CDMA is asymptotically seven times larger as compared to the capacity of TDMA. In [7], where some practical non-military applications of spread spectrum were listed, an estimate of the CDMA radio capacity indicates a possible advantage of CDMA over TDMA by a factor of three to six. However, most of the results in favor of one or the other technique were based on incomplete models. As by now, the only proven advantage of CDMA that has no appropriate counterpart in TDMA is the possibility of overlaying high processing gain spread spectrum users over existing narrow-band users [8], [9]. The impact of different techniques and technologies on TDMA and CDMA will be discussed in the following sections. It is our intention to offer an answer to the question which of the two multiple access techniques, TDMA or CDMA, is more appropriate for application in the future PCN.

Since the largest concentrations of PCN users are expected in densely populated urban areas, a microcellular environment is of most interest for studying the radio capacity issues of PCN. In other words, it is the sector of personal mobile communications where the maximum radio capacity should be achieved and, therefore, we will concentrate on the microcellular layout.

One of the most often reported advantages of CDMA over TDMA is the capacity improvement through the interference reduction [10]. This is due to the fact that CDMA is multiple-access interference (MAI) limited, while TDMA is bandwidth limited and, therefore, a reduction in MAI contributes to an increase of the number of simultaneous spread spectrum users.[1] Although true in principle, such a favoring of CDMA is not quite fair for the following reason. In general, the performance of TDMA is intersymbol interference limited or, in the case of a good equalization scheme, bandwidth limited, given that the co-channel separation is large. However, a large co-channel distance separation of TDMA suffers from a low spectral efficiency of TDMA. Any measure that would reduce the effect of co-channel interference would allow the use of a smaller cluster size and, hence, improve the radio capacity of TDMA. For example, the use of cell sectorization and/or adaptive antenna arrays may reduce the cluster size in TDMA to 1, in the limit [26]. That is, a manifold increase in the TDMA radio capacity is possible. Thus, a reduction in the co-channel interference (CCI) would improve the radio capacity of both, TDMA and CDMA. The question is only in the practical achievability of the limiting capacity.

For example, the use of forward error correction (FEC) is often seen as more advantageous in CDMA than in TDMA. Indeed, FEC can improve the radio capacity of CDMA by a factor of about 2.5 times, with no bandwidth expansion [11]. On the other hand, the use of FEC in TDMA requires a bandwidth expansion in order to keep a fixed data rate, which may result in a spectrum

---

[1] In evaluating the performance of CDMA, a simple correlation receiver is usually considered, although an optimum multiuser detector can closely approach the single-user performance, with additional software complexity only [24].

efficiency degradation. However, if applied properly, FEC can actually improve the spectrum efficiency of TDMA also. For example, by using a high rate code, say r=3/4, with a coding gain of 4 dB, the system can stand a larger CCI and, hence, the cluster size can be reduced. More specifically, by using such a code, the cluster size can be reduced from 7 to 4 without a degradation in performance [12]. In the European GSM system [27], the cluster sizes, with FEC, of 4 and 3 are anticipated as pessimistic and optimistic, respectively. The effective improvement in the spectrum efficiency is more than 1.3 times, which is significantly smaller that in the CDMA case. However, by using a trellis coded modulation (TCM) scheme, a coding gain of more than five dB can be achieved with no bandwidth penalty. TCM in combination with some co-channel interference reduction technique can allow the use of the smallest cluster size, which would contribute to a significant improvement of the TDMA radio capacity. Indeed, TCM is seen as a means of achieving improved spectrum efficiency of cellular communications [14].

It was reported that the CDMA radio capacity actually doubles in the case of voice transmission, due to the voice activity factor of 40% for one-way direction of a communication link [8], [10]. That is, during a silence period the channel can be used by some other user. This property is exploited in CDMA for capacity improvement without significant additional complexity in the design. It is only required to suppress the carrier during the silence intervals. A similar improvement is possible to achieve in TDMA as well. An estimate of the radio capacity improvement in TDMA using packet reservation multiple access is given in [15]. Although in the base-to-mobile direction statistical multiplexing is straightforward, in the opposite direction some overhead is needed for the channel access control. It was found that 37 simultaneous speech transmissions can be maintained with 20 slots per frame, with a packet dropping probability of less than 1%. Hence, this is only slightly worse compared to what is achievable in CDMA. In other words, even in the case of voice transmission, CDMA is not as favorable as it was until recently taught.

The behavior of both, CDMA and TDMA in fading and multipath environments was thoroughly studied over years. It is generally agreed that CDMA is less sensitive to these phenomena than TDMA [10], [11]. In addition, CDMA can use multipath propagation to its advantage by using the RAKE principle [16]. On the other hand, there are several equalization techniques at the designer's disposal for combating the multipath induced impairments [17], [18]. The most critical aspect of equalization in time-variant channels is the equalizer ability to track the changes of the channel impulse response [17], [19], [20]. The two competing approaches are the block and sequential equalizer operations and a further investigation is necessary to answer the question as to which of the two options is more suitable for different PCN environments [17]. Also, an extension of the results in [20] to include the bit-error-rate (BER) performance with imperfect channel impulse response estimates would be very useful in evaluating the performance of TDMA for PCN. In mobile radio channels with slower rates of changes, e.g.. picocellular and microcellular, a transmitter based equalization scheme may be more attractive because of a smaller computational burden on the mobile terminal. In particular, trellis precoding [21], that is, combined coding, precoding and shaping can be used to approach closely the channel capacity, even in the intersymbol interference channels. Since an estimate of the channel impulse response is required for precoding in the transmitter, either the receiver has to supply that information to the transmitter or, assuming that the channel is reciprocal, a transmitter based channel estimation scheme may suffice. Since the ambient noise may be quite different at the transmitter and receiver, the later approach may require a method for channel impulse response estimation which has low sensitivity to the additive white Gaussian noise. That is, a polyspectra method for channel estimation may be desirable [22], but with an increased computational burden. As alternatives to trellis precoding, various forms of multicarrier modulation [23] may be attractive and feasible with the current state of the art in the DSP technology. Finally, the role of diversity in the performance improvement of TDMA and CDMA is well accepted [18], [25]. Still, a more elaborate investigation of the diversity for cellular CDMA with coding, for a frequency selective channel, is needed [25], as well as appropriate comparison with the diversity effects in TDMA.

Having identified the most important disputes and points of interest for a comparison of TDMA and CDMA for PCN, some of the concepts discussed will be analyzed in a comparison example in the next section. The purpose of this example is to demonstrate how different techniques may affect the capacity of TDMA and CDMA and, correspondingly, how conclusions regarding the choice of one or the other technique may be context dependent.

PERFORMANCE COMPARISON OF TDMA AND CDMA

Channel Model and Microcellular Layout

Microcellular radio channels are characterized by a large number of reflections, inherent to urban radio-wave propagation conditions. Correspondingly, the actual discrete multipath channel can be well approximated with a continuous multipath fading model [29]. The measurements of microcellular channel impulse response indicate a systematic presence of a strong specular component. Typically, the ratio of random to specular signal power is -7 dB as indicated in [28]. Hence, we assume the equivalent low-pass impulse response to be a superposition of a Dirac delta function, corresponding to the specular component, and a zero-mean complex Gaussian random process $\beta(\tau)$ representing the time-dispersive signal components. $\beta(\tau)$ is wide-sense-stationary with the following properties:

$$E[\beta(\tau_1)\beta(\tau_2)] = 0 \tag{1}$$

$$E[\beta(\tau_1)\beta^*(\tau_2)] = 2\zeta(\tau_1)\delta(\tau_1 - (\tau_2)) \tag{2}$$

This model is usually referred to as a wide sense stationary uncorrelated scattering (WSSUS) channel. $\zeta(\tau)$ is most often called multipath intensity profile and represents the power-delay distribution of the multipath channel. Since our results are insensitive to the actual multipath profile of a given duration, we assume, for simplicity, a rectangular multipath intensity profile defined as,

$$\zeta(\tau) = \frac{1}{2T_0}, \text{ for } |\tau| \le T_m \tag{3}$$

where T_m represents two-sided maximum dispersion and T_0 is an arbitrary scaling factor. In order to get specific results we assume that T_0 corresponds to the symbol duration T. For non-diversity like modulation schemes the actual shape of the multipath intensity profile is unimportant. The root mean square value of multipath spread, which is in this particular case found as $t_{rms} = T_m/\sqrt{3}$, is the principal measure. In the case of equalization, the shape of the multipath intensity profile does not affect performance as long as the equalizer span exceeds $2T_m$. When $2T_m$ is larger than the equalizer span, better performance is achieved for multipath intensity profiles with higher multipath power percentages within the equalizer span. The performance of a TDMA system, which is sensitive to the instantaneous interference, is dominantly determined by the maximum value of r.m.s. multipath spread, which is about 3μsec for urban microcellular channels [30].

In order to analyze the affect of co-channel interference, we consider the worst-case mobile position at the cell edge. To get specific results, a hexagonal microcellular layout is considered. The co-channel interference power is added incoherently using a Gaussian assumption, which is actually improved in the presence of multipath propagation. From the geometry of a seven cell two-dimensional cluster we find that median co-channel interference power is proportional to [14]:

$$I \propto 2[(D-R)^{-\gamma} + D^{-\gamma} + (D+R)^{-\gamma}] = F_c R^{-\gamma} \tag{4}$$

with $F_c = 0.033$. D and R correspond to the co-channel distance separation and cell radius, respectively. We assume the path loss exponent $\gamma = 3.6$.

Cell sectorization can improve the radio capacity of TDMA through a reduction of co-channel interference. In order to quantitatively explore this effect we consider cluster size K=3 with 120° sectoral antennas characterized with the front-to-back radiation ratio of 10 dB. With these parameters we find:

$$\begin{aligned}
I \propto\ & D^{-\gamma} + (D+0.7R)^{-\gamma} + 0.1\,[(D-R)^{-\gamma} \\
& + (D+R)^{-\gamma} + (D-0.3R)^{-\gamma} + (D+0.7R)^{-\gamma}] \\
& = F_c' R^{-\gamma}
\end{aligned} \tag{5}$$

where $F'_c=0.0417$, which corresponds to only 1 dB degradation of carrier-to-interference ratio as compared to cluster size 7 with omnidirectional antennas.

TDMA Performance Analysis

The signal transmitted by the base station i is defined as

$$s_i(t) = \text{Re}\left[Ad_i(t)e^{j(\omega_0 t+\theta_i)}\right] \tag{6}$$

where $d_i(t)$ represents the quadrature phase shift keying (QPSK) signal defined as

$$d_i(t) = A \sum_{j=-\infty}^{\infty} s_i(j)p(t - jT) \tag{7}$$

where $1/T$ corresponds to the TDMA gross symbol rate. For k TDMA users with data rates $1/T_b$ we have $T=2T_b/k(1+p)$, where p corresponds to the TDMA overhead relative to the information symbols. The complex QPSK symbols, $s_i(j)$, take on values $(\pm 1 \pm j)$.
p(t) is a rectangular pulse of unit amplitude and duration T.

The equivalent baseband received signal at the desired mobile, from the i-th base station, is given by:

$$r_i(t) = [\alpha_i d_i(t-\tau_i) + \xi_i A d_i(t-\tau_i) * \beta_i(t)] e^{j\theta_i} \tag{8}$$

where τ_i and θ_i represent random delay and phase of the i-th base station signal. The symbol * denotes convolution. As a convention we assign i=0 to the desired base station signal and assume perfect timing and phase synchronization, that is $t_0=0$ and $\theta_0=0$. The total received signal at a mobile in the microcell i=0 is

$$r(t) = r_0(t) + \sum_{i=1}^{6} r_i(t) + n(t) \tag{9}$$

where $n(t)$ is additive Gaussian noise with one-sided power spectral density N_0. The second term in (9) represents the total co-channel interference.

For the performance of equalized TDMA we consider a decision feedback equalizer. The forward filter consists of $k=1+k_1+k_2$ taps with $\tau_s=T/2$ spacing, where k_1 and k_2 correspond to "future" and "past" taps, respectively. The forward filter is followed by a matched filter. Assuming that the feedback filter cancels the intersymbol interference from previously detected symbols, the decision variable at the sampling instant t_0 is given by

$$
\begin{aligned}
z = &\sum_{m=k_1}^{k_2} w_m^* \{\alpha_0 g_0 (t_0 - m\tau_s) + \xi_0 q_0 (t_0 - m\tau_s) \\
&+ \sum_{i=1}^{6} [\alpha_i g_i(t_0\text{-}\tau_i\text{-}m\tau_s)+\xi_i q_i(t_0\text{-}\tau_i\text{-}m\tau_s)]e^{j\theta_i} \\
&+ v(t_0 - m\tau_s)\}
\end{aligned}
\tag{10}
$$

where w_m's represent the forward filter tap coefficients and

$$
\begin{aligned}
g(t) &= d(t) * p(-t) \\
q(t) &= g(t) * \beta(t) \\
v(t) &= n(t) * p(-t)
\end{aligned}
\tag{11}
$$

From (10) we find the total interference power as

$$
\rho^2 = \frac{N_0}{T} \underline{w}^{*T} \underline{G} \underline{w}
\tag{12}
$$

where $\underline{G}$ is a k x k dimensional matrix representing the contribution of all interfering signals. From (10) and (12) and by using the Schwartz Inequality, for $\underline{w}_{opt}=\underline{G}^{-1} \underline{b}$, we get

$$
S N R_{max} = a\underline{b}^{*T} \underline{G}^{-1} \underline{b}
\tag{13}
$$

where $\alpha=(\alpha A)^2 T/2N_0$, and to get specific results we assume that all signals experience the same multipath channel. The k dimensional vector $\underline{b}$ corresponds to useful signal components at the output of the forward filter. The matrix $\underline{G}$ is conditioned on co-channel interference symbols, channel impulse responses $\beta_i(\tau)$, delays τ_i and random phases θ_i. The calculation of BER requires averaging of the quadratic form in (13) which is composed of a random vector and a random matrix. Following the approach in [31] we first average the matrix $\underline{G}$ to get an estimate of BER given by

$$
P_d = 0.5 \prod_{m=k_1}^{k_2} (1 + a\lambda_m)^{-1}
\tag{14}
$$

where λ_m's are the eigenvalues of the matrix

$$\underline{B} = E\left[\underline{G}^{-1/2}\,\overline{\underline{b}\,\underline{b}^{*T}}\,\underline{G}^{-1/2}\right] \tag{15}$$

For ordinary QPSK with matched filter detection, the decision variable is given by

$$\begin{aligned}
z' &= \alpha_0 g_0(t_0) + \xi_0 q_0(t_0) + v(t_0) \\
&+ \sum_{i=1}^{6} [\alpha_i g_i(t_0 - \tau_i) + \xi_i q_i(t_0 - \tau_i)]\,e^{j\theta_i}
\end{aligned} \tag{16}$$

Because of quadrature symmetry, only the real part of z' may be considered. The intersymbol interference (ISI) and cross-rail interference (CRI) at the output of the matched filter receiver are zero-mean complex Gaussian random variables. The variances of ISI and CRI, conditioned on the data sequence, are found as

$$\sigma^2 = (\xi A T)^2 [I_1 + (s_0 s_1 + s_0 s_{-1})\,I_2] \tag{17}$$

where

$$\begin{aligned}
I_1 &= S'\left(1 - S' + \frac{2}{3}S'^2\right) \\
I_2 &= S'^2\left(\frac{1}{2} - \frac{S'}{3}\right)
\end{aligned}$$

where $S' = T_m/T$. In the case of ISI, s corresponds to a data bit in the in-phase branch and, similarly, to a data bit in the quadrature branch for the CRI. Assuming that the co-channel interference (CCI) is a zero-mean Gaussian random variable, the variance of CCI is found, averaged over random phases, delays and bit sequences, as

$$\sigma_{CCI}^2 = \frac{(AT)^2 F_c}{3}\left[2\alpha^2 + 6\xi^2 I_1\right] \tag{18}$$

Using the results from [12], with (17) and (18), we find the average in-phase SNR as

$$\begin{aligned}
[S\,N\,R(\underline{s})]^{-1} &= y[I_1 + (s_{I,0}s_{I,1} + s_{I,0}s_{I,-1})\,I_2] \\
&+ y[I_1 + (s_{Q,0}s_{Q,1} + s_{Q,0}s_{Q,-1})I_2] \\
&+ \frac{1}{2a} + F_c\left(\frac{2}{3} + 2yI_1\right)
\end{aligned} \tag{19}$$

where $y = \xi^2/\alpha^2$ corresponds to the ratio of random to specular signal power. Finally, assuming Gray encoding of QPSK symbols, the BER for QPSK is approximately

$$P_{nl} = \overline{Q((\sqrt{S\,N\,R_{(s)}}))} \tag{20}$$

where averaging is over all possible combinations of data bits. The inflection point of P_{nl} is at $P_{nl}=4 \times 10^{-2}$, where P_{nl} is convex $\cup$ for smaller values. Hence, averaging of the second moment of co-channel interference as obtained in (18), leads to a lower bound on (20), for value of P_{nl} of practical interest. However, since nonequalized QPSK is ISI and CRI limited, (20) represents a close approximation to true BER for all practical purposes.

CDMA Performance Analysis

In CDMA, the signal transmitted by the base station i is defined by:

$$s_i(t) = \mathrm{Re}\left[Ab_i(t)\,d_i(t)\,e^{j(\omega_0 t+\theta_i)}\right] \tag{21}$$

where $b_i(t)$ represents the spreading waveform composed of T_c long rectangular chips with random independent identically distributed equiprobable amplitudes ± 1. All other parameters are defined as in (6). Using the results from [32], the variances of ISI and CRI are given by:

$$\sigma^2 = \frac{2(\xi AT)^2}{3N}\left(\frac{1}{2} + \frac{L}{N}\right) \tag{22}$$

where $N=T/T_c$ is the processing gain and L is defined by $T_m = LT_c \leq T$. Similarly, the variance of MAI is found as

$$\sigma_I^2 = \frac{2(\alpha AT)^2}{3N}\left(1 + 4y\,\frac{L}{N}\right)(KF_s - 1) \tag{23}$$

where K represents the number of active users in every cell and $F_s = 3.74$, considering twelve interfering cells around the worst case mobile position. Since we are interested in a large number of users, the Gaussian approximation for MAI is appropriate [34], and the BER is then given by:

$$P_b \approx Q\left(\sqrt{S\,N\,R_s}\right) \tag{24}$$

where

$$\overline{S\,N\,R_s}^{-1} = \frac{1}{2a} + \frac{2y}{3N}\left(1 + 2\,\frac{L}{N}\right) + \frac{2(K\,F_s - 1)}{3N}\left(1 + 4y\,\frac{L}{N}\right) \tag{25}$$

Coded Performance

For the illustration of coded performance, we consider several examples of convolutional codes. The probability of error of Viterbi decoding for a code of rate $r_c = b/V$ is upper bounded by [35], [36]:

$$P_b \leq \frac{1}{b} \sum_{n=d_{free}}^{\infty} c_n \, P_{2n} \tag{26}$$

where P_{2n} represents the pairwise error probability, which is the probability that the wrong path at the distance n is selected. The coefficients c_n correspond to the number of bit errors in adversaries at the distance n. For the best rate 1/2 convolutional code and punctured convolutional codes of rates 3/4 and 7/8, c_n's are given in [37] and [36], respectively.

Since we assumed Gaussian interference in CDMA and equalized TDMA, the pairwise error probability for CDMA is given by:

$$P_{2n} = Q\left(\sqrt{nS \, N \, R_{sc}}\right) \tag{27}$$

and for equalized TDMA it is derived as:

$$P_{2n} = 0.5 \prod_{l=-k_1}^{k_2} \left(1 + n^2 a_c \lambda_l\right)^{-1} \tag{28}$$

where the subscript c in both cases indicates that the corresponding expressions should be adjusted to the code rate.

In the case of nonequalized TDMA, the derivation of the pairwise error probability is slightly more complicated. First, it should be noted that the ISI and CRI are independent with identical statistics with variances given by (17). The sum of ISI and CRI variances can be written as:

$$\sigma_t^2 = F_1 + F_2 f(s_I, s_Q) \tag{29}$$

where F_1 and F_2 are functions of the pulse shape and fading channel, and can be assumed constant for slow fading. $f(s_I, s_Q)$ is a discrete valued function representing possible combinations of in-phase and quadrature symbols. For the simplicity of implementation, we assume soft decision decoding without channel-state information. The maximum likelihood metric in this case is given by [39]:

$$m(z_k', d_k) = z_k' d_k \tag{30}$$

where z' is defined by (16) and d_k represents the k-th coded symbol in the coded sequence. The discrete valued function $f_I(\underline{s_I},\underline{s_Q})$ takes on five possible values, 0, ±2 and ±4, with probabilities 3/8, 1/4 and 1/16, respectively. Correspondingly, σ^2 in (29) can take five values σ_1^2, ..., σ_5^2, with probabilities p_1, ..., p_5. Finally, the pairwise error probability is found as:

$$P_{2n} = \sum_{k_1=0}^{n} \binom{n}{k_1} p_1^{k_1} \sum_{k_2=0}^{n-k_1} \binom{n-k_1}{k_2} p_2^{k_2} \cdots \sum_{k_4=0}^{n-\sum_{j=1}^{3} k_j} \binom{n - \sum_{j=1}^{3} k_j}{k_4} p_4^{k_4}$$

$$\times p_5^{n-\sum_{j=1}^{4} k_j} P_{2n}(k_1, ..., k_5)$$

(31)

where $P_{2n}(k_1, ..., k_5)$ represents the conditional pairwise probability of error defined by:

$$P_{2n}(k_1, ..., k_5) = Q\left[n\alpha A T_s \left(n(N_0 T_s + \sigma_{CCI}^2) + \sum_{j=1}^{5} k_j \sigma_j^2 \right)^{-1/2} \right]$$

(32)

where T_s represents the coded bit duration.

NUMERICAL RESULTS

The performance of equalized and non-equalized TDMA is compared in Figure 1. One-sided multipath delay dispersion of 5μsec and a rectangular multipath intensity profile are assumed. The channel is Rician and the ratio of specular to multipath power is -7dB. The equalizer performance represents the steady-state condition and the channel is assumed to be time-invariant. An increase in the number of users per TDMA carrier, for a fixed data rate of 9.6 Kbps, corresponds to an increase of the normalized multipath spread. The fractional overhead per user is p=15%. Not surprisingly, the performance of equalized TDMA improves for larger equalizer lengths, as the normalized multipath spread increases, which agrees with the conclusion in [29]. This is due to an implicit diversity improvement achieved by equalization. However, in a dynamic fading channel, a smaller number of equalizer taps may be desirable because of the excess mean-square error in dynamically tracking the channel impulse response. Hence, the results obtained for a 7-tap equalizer may be too optimistic for an equalized TDMA system. Regardless of that, however, in the region of most interest for comparison of equalized and non-equalized TDMA (bellow about 200 Kbps aggregate TDMA data rates), all the equalizers considered lengths give approximately the same performance. Hence, the equalizer length will not affect the generality of the conclusions in comparing equalized and non-equalized TDMA. Clearly, from Figure 1, it can be concluded that non-equalized TDMA is a preferable technique for TDMA data rates bellow about 100 Kbps, for this specific channel. Again, this is an optimistic estimate for equalized TDMA, because of an additional performance degradation which can be expected in a dynamic fading channel.

Comparing performance behaviors in Figure 1 and Figure 2, it can be seen that with a reduction in the signal-to-thermal noise ratio from 20dB to 10dB, the advantage of non-equalized TDMA becomes more significant. In Figure 2 the dashed line labeled as "genie" represents the performance of TDMA with intersymbol interference completely removed. The genie curve doesn't account for additional improvements due to implicit diversity that can be achieved by using an equalization scheme. Naturally, the genie performance is uniformly better than non-equalized TDMA. What may not be clear a priori, is that the genie is uniformly better than equalized TDMA as well. This implies that the contribution of residual intersymbol interference and noise enhancement is more dominant than implicit diversity improvement of the decision feedback equalizer.

In Figure 3 the performance of CDMA and TDMA is compared. In both cases the total system bandwidth is same as determined by TDMA system requirements to support a given number of active users. The capacity improvement due to voice activity by a factor of 2.25 is assumed in the CDMA case, corresponding to a voice activity factor of 40%. This is a slightly optimistic estimate of capacity improvement and represents an upper bound [9]. In the TDMA case a voice activity capacity improvement factor of 1.85 is assumed even though it is more difficult to achieve than in CDMA. Again, it is an optimistic estimate because a finite packet dropping probability to achieve this statistical multiplexing gain is neglected [16]. Apart from these approximations, it is clear that CDMA can outperform TDMA for certain ranges of number of users or, equivalently, for a given range of system bandwidths. More specifically, CDMA will outperform TDMA for more than 47 active users per cell. On the other hand equalized TDMA is uniformly better than CDMA, unless other techniques are employed such as RAKE, coding, sectorization, etc. One should be careful in interpreting these results, however. It is assumed that all TDMA users in a cell are multiplexed on a single TDMA carrier. Hence, one can conclude that a hybrid FDMA/TDMA multiple access scheme, with data rates per TDMA carrier bellow about 450 Kbps will always perform better than CDMA. These results are of importance only if FEC is considered and it is well understood that FEC must be an integral part of a CDMA system. Namely, FEC can improve the radio capacity of a CDMA system by a factor of 2-3, without a bandwidth penalty [12]. Nevertheless, this result will represent a baseline performance and serve the useful purpose in demonstrating how additional coding and signal processing techniques will alter this conclusion.

Figure 4 demonstrates the effect of using sectoral 120° antennas, with all other parameters same as in Figure 3. Although the performance of both access methods improves, one should note significantly greater improvement in the CDMA case, relative to TDMA. This is not surprising because CDMA operates in a mostly multiple access interference limited region. For a given, fixed cluster size, the TDMA radio capacity can not improve, while any interference reduction in CDMA results in a capacity improvement. Although we are interested in the radio capacity comparison of TDMA and CDMA, the number of users versus BER is shown in all these figures. However, relative improvements in the CDMA radio capacity can be evaluated by noting (24)-(25), by which the BER is related to the number of active users. It should be noted that equalized TDMA is not uniformly better than CDMA when sectorization is included. Indeed, CDMA will outperform both non-equalized and equalized TDMA for data rates in excess of 140 Kbps and 310 Kbps, respectively.

A similar qualitative behavior of the multiple access schemes can be deduced from Figure 5, where FEC is used instead of sectorization. In the CDMA case a rate 1/3 constraint length 7 convolutional code is used. In the TDMA case a trade off between coding gain and spectrum efficiency is achieved with a rate 3/4, punctured convolutional code of constraint length 7. Comparing the results in Figures 4 and 5 with those in Figure 3, it can be concluded that significant performance improvement, in terms of the BER, can be used to allow for a higher level of co-channel or multiple-access interference. Consequently, in Figure 6 we analyze the effect of reduced cluster size, M=3. It can be seen that both equalized and non-equalized TDMA can operate with this cluster size when FEC is used. Both, absolute and relative performance of CDMA are degraded as compared to the results in Figures 4 and 5. This is a direct consequence of reduced total system bandwidth, which results in a reduced CDMA processing gain. Hence, the flaw in the argument that CDMA is interference limited while TDMA is bandwidth limited is obvious in this example. More precisely, TDMA is truly bandwidth limited only when the smallest cluster size M=1 is used. We speculate that this cluster size can be achieved in the limit, by using additional signal processing techniques for reduction of co-channel interference or mitigation of its effects. By reducing the cluster size the spectrum efficiency of TDMA improves. On the other hand, for a fixed cluster size a reduction in co-channel interference can improve the BER but not the spectrum efficiency as in the CDMA case. These effects can be seen from the performance curves presented in Figures 6 and 7.

Finally, in Figure 8 the limiting cluster size M=1 is considered. FEC and 60° sectoral antennas are assumed. This corresponds to the bandlimiting scenario for TDMA and any further reduction in multiple access interference or signal processing gain will only improve the radio capacity of CDMA. There are several techniques by which this effect can be achieved. Multi-user detection, diversity and adaptive beamforming are already identified as potential signal processing techniques for further CDMA radio capacity improvement. It should be noted, however, that previous comparisons of equalized TDMA and CDMA were not quite fair to CDMA. It would be more appropriate to compare equalized TDMA with CDMA employing the RAKE receiver, to account for implicit diversity improvement of equalization. Regardless of that, and the fact that some potential system impairments were neglected, like flat fading, frequency dispersion, imperfect power control and synchronization, this comparison presented in a chronological order clearly demonstrates the importance of a comprehensive radio capacity comparison for PSN multiple access. The omission of some system impairments or feasible signal processing techniques can lead us to a suboptimal solution for future PCN.

CONCLUSIONS

It was shown in this paper that the capacity of both CDMA and TDMA can be significantly improved by applying techniques which reduce co-channel interference, for example cell

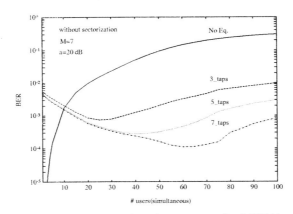

Figure 1. Effect of number of taps on equalized TDMA.

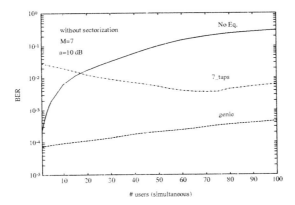

Figure 2. Genie TDMA performance.

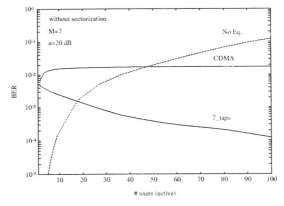

Figure 3. Comparison of TDMA and CDMA with voice activity.

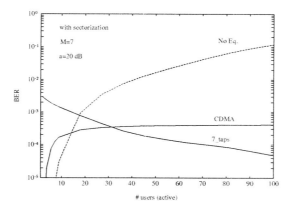

Figure 4. Effect of sectorization on TDMA and CDMA.

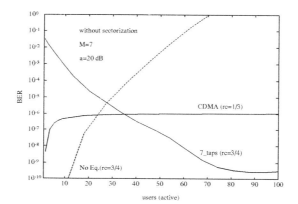

Figure 5. Effect of FEC on TDMA and CDMA.

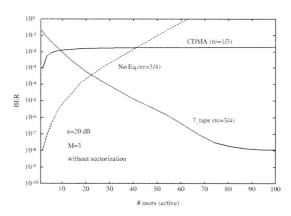

Figure 6. TDMA and CDMA performance for cluster size 3.

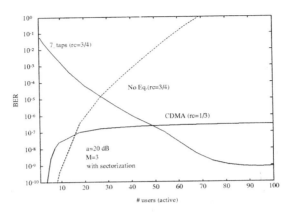

Figure 7. Combined effect of FEC and sectorization.

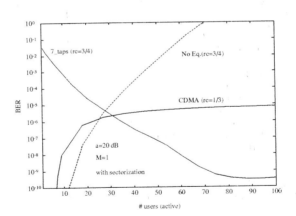

Figure 8. TDMA and CDMA performance for cluster size 1.

sectorization. Numerical results clearly indicate that there is no simple answer to the question as to which of the two techniques yields larger capacity. Nevertheless, one can conclude that, in the limit, when TDMA becomes fully bandwidth limited, CDMA can further improve capacity by applying some additional interference reduction techniques or enable it to operate at a lower SNR and, ultimately, outperform TDMA.

The comparison of equalized and unequalized TDMA reveals that in a Rician channel at relatively small and moderate values of normalized multipath spread unequalized TDMA performs better than equalized TDMA.

REFERENCES

[1] R.L. Pickholtz, D.L. Schilling and L.B. Milstein, "Theory of spread spectrum - a tutorial," IEEE Trans. on Comm., vol. COM-30, NO. 5, Jan. 1989.

[2] D. Muilwijk, "On the spectrum efficiency of radio transmission schemes for cellular radio", LINKS FOR THE FUTURE, IEEE/Elsevier Science Publishers B.V. (North-Holland), 1984.

[3] A.J. Viterbi,"When not to spread spectrum - a sequel," IEEE Comm. Mag., vol. 23, NO. 4, Apr. 1985.

[4] C.Y. Lee,"Spectrum efficiency in cellular", IEEE Trans. on Veh. Tech., vol. VT-38, NO. 2, May 1989.

[5] On-Ching Yue, "Spread spectrum mobile radio, 1977-1982", IEEE Trans. on Veh. Tech., vol. VT-32, NO. 1, Feb. 1983.

[6] G.R. Cooper and R.W. Nettleton, "A spread spectrum technique for high-capacity mobile communications," IEEE Trans. on Veh. Tech., vol. VT-27, NO. 4, Nov. 1984.

[7] K.S. Gilhousen, I.M. Jacobs, R. Padovani and L.A. Weaver, Jr., "Increased capacity using CDMA for mobile satelite communications," IEEE J. on Sel. Ar. in Comm., vol. JSAC-8, NO. 4, May 1989.

[8] D.J. Schilling, R.L. Pickholtz and L.B. Milstein, "Spread spectrum goes commercial," IEEE Spectrum, Aug. 1990.

[9] R.L. Pickholtz, L.B. Milstein and D.J. Schilling, "Spread spectrum for mobile communications," IEEE Trans. on Veh. Tech., vol. VT-40, NO. 2, May 1991.

[10] L.B. Milstein et al.,"On the feasibility of a CDMA overlay for personal communication networks," IEEE J. on Sel. Ar. in Comm., vol. SAC-10, NO. 4, May 1992.

[11] K.S. Gilhousen, I.M. Jacobs, R. Padovani, A.J. Viterbi, L.A. Weaver and C.E. Wheatley, "On the capacity of a cellular CDMA system," IEEE Trans. on Veh. Tech., vol. VT-40, NO. 2, May 1991.

[12] B.R. Vojcic, R.L. Pickholtz and I.S. Stojanovic, "A comparison of TDMA and CDMA in microcellular radio channels," Proceeding of the IEEE ICC'91, Denver, 1991.

[13] B.R. Vojcic, R.L. Pickholtz and A.S. Ragab, "TDMA Performance for Personal Communications Networks", Communication Theory Mini Conference (in conjunction with GLOBECOM '92), 1992.

[14] W.L. Lee, Mobile Cellular Telecommunications Systems, McGraw Hill, 1988.

[15] K. Feher, "MODEMS for emerging digital cellular-mobile radio systems," IEEE Trans. on Veh. Tech., vol. VT-40, NO. 2, May 1991.

[16] S. Nanda, D.J. Goodman and U. Timor, "Performance of PRMA: A packet voice protocol for cellular systems," IEEE Trans. on Veh. Tech., vol. VT-40, NO. 3, Aug. 1991.

[17] G.L. Stuber and C. Kchao, " Analysis of a multiple-cell direct-sequence CDMA cellular mobile radio system," IEEE J. on Sel. Ar. in Comm., vol. SAC-10, NO. 4, May 1992.

[18] J.G. Proakis, "Adaptive equalization for TDMA mobile radio," IEEE Trans. on Veh. Tech., vol. VT-40, NO. 2, May 1991.

[19] P. Balaban and J. Salz, "Optimum diversity combining and equalization in digital data transmission with applications to cellular mobile radio - Part I and II," IEEE Trans. on Comm., vol. COM-40, NO. 5, May 1992.

[20] R.A. Ziegler and J.M. Cioffi, "Estimation of time-varying digital radio channels," IEEE Trans. on Veh. Tech., vol. VT-41, NO. 2, May 1992.

[21] N.W.K. Lo, D.D. Falconer and A.U.H. Sheikh, "Adaptive equalization and diversity combining   for mobile radio using interpolated channel estimates," IEEE Trans. on Veh.Tech., vol. VT-40, NO. 3, Aug. 1991.

[22] M.V. Eyuboglu and G.D. Forney, Jr., "Trellis precoding: combined coding, precoding and shaping for intersymbol interference channels," IEEE Trans. on Info. Th., vol. IT-38, NO. 2, March 1992.

[23] D. Hatzinakos and C.L. Nikias, "Estimation of multipath channel response in frequency selective channels," IEEE J. on Sel. Ar. in Comm., vol. SAC-7, NO. 1, Jan. 1989.

[24] J. Bingham, "Multicarrier modulation for data transmission: An idea whose time has come,"IEEE Comm. Mag., vol. 28, NO. 5, May 1990.

[25] H.V. Poor, "Signal processing for wideband communications," IEEE Info. Th. Soc. Newsletter,vol. 42, NO. 2, June 1992.

[26] L.B. Milstein, T.S. Rappaport and R. Barghouti, "Performance evaluation for cellular CDMA," IEEE J.on Sel. Ar. in Comm., vol. SAC-10, NO. 4, May 1992.

[27] S.C. Swales, M.A. Beach, D.J. Edvards and J.P. McGeehan, "The performance enhancement of multibeam adaptive base-station antennas for cellular land mobile radio systems," IEEE Trans. on Veh. Tech., vol. VT-39, NO. 1, Feb. 1990.

[28] K. Raith and J. Uddenfeldth, "Capacity of digital cellular TDMA systems," IEEE Trans. on Veh. Tech., vol. VT-40, NO. 2, May 1991.

[29] T. Sexton and R. Pahlavan, "Channel modeling and adaptive equalization for indoor radio channels," IEEE J. on Sel. Areas in Comm., vol. SAC-7, NO. 1, Jan. 1990.

[30] R. Bultitude and G. Bedal, "Propagation characteristics on microcellular urban radio channels," IEEE J.on Sel. Areas in Comm., vol. SAC-7, NO. 1, Jan 1989.

[31] P. Monsen, "Theoretical and measured performance of a DFE modem on a fading multipath channel," IEEE Trans. on Comm., vol. COM-25, NO. 10, Oct. 1977.

[32] B.R. Vojcic and R.L. Pickholtz, "Performance of coded direct sequence spread spectrum in a fading dispersive channel with pulsed jamming," IEEE J. on Sel. Areas in Comm., vol. SAC-8, NO. 5, June 1990.

[33] D.E. Borth and M.B. Pursley, "Analysis of direct sequence spread spectrum multiple access Communications over Rician fading channels," IEEE Trans. on Comm., vol. COM-27, NO. 10, Oct. 1979.

[34] R.K. Morrow, Jr., and J.S. Lehnert, "Bit-to-bit error dependence in slotted DS/SSMA packet systems with random signature sequences," IEEE Trans. on Comm., vol. COM-37, NO. 10, Oct. 1989.

[35] A.J. Viterbi, "Convolutional codes and their performance in communication systems," IEEE Trans.on Comm., vol. COM-19, Oct. 1971.

[36] S. Kallel and D. Haccoun, "Generalized type II hybrid ARQ scheme using punctured convolutional coding," vol. COM-38, NO. 11, Nov. 1990.

[37] J. Conan, "The weight spectra of some short low-rate convolutional codes," IEEE Trans. on Comm., vol. COM-32, NO. 9, Sept. 1984.

[38] B.R. Vojcic, "Performance of a class of digital transmission shemes in a fading dispersive channel with jamming," PhD thesis (in serbo-croatian), Faculty of Electrical Engineering, Belgrade University, Belgrade 1989.

[39] M.K. Simon et al., Spread Spectrum Communications, Computer Science Press, 1985.

THE MILLICOM/SCS MOBILECOM PCN FIELD TEST

Donald L. Schilling
Dept. of EE
CCNY
New York,NY 10031

Laurence B. Milstein
Dept. of ECE
U. of CA. San Diego
La Jolla, CA 92093

Raymond L. Pickholtz
Dept. of EECS
George Washington U.
Washington, DC 20052

Marvin Kullback
SCS Mobilecom, Inc.
Port Washington, NY 11050

William Biederman
Millicom, Inc.
New York, NY 10022

Code Division Multiple Access (CDMA) is a digital multiple-access technique, whereby each signal has its own unique binary sequence, and all signals share the same spectrum. In the United States, personal communication networks using CDMA, will provide to each user, a separate code, which will change from cell-to-cell. We call this code reuse, just like the frequency reuse employed in the FDMA or TDMA systems. However, there are many more codes, than frequency bands.

CDMA provides total transmission privacy, insures high quality voice, high data rate transmission and no outage time. Applications of Broadband CDMA include Personal Communications, Wireless PBX, Point-to-Point microwave, Wireless Local Area Nets, etc.

In the United States, the FCC has designated a Personal Communication band of 1800-2000 MHz. In addition to the PCN band, there are three, Part 15, unlicensed spread spectrum bands. Many of our patents can be employed on these bands.

Broadband CDMA is a spread spectrum technique, which produces extremely high quality communications, similar to PSTN quality. High voice quality is achieved, using adaptive data modulation at 32 kilobits per second. FAX and modem interfaces are provided. Extremely low outage time results, since the broadband spread spectrum is not affected by fading. Indeed the fade margin, is typically 1 to 2 dB. Extremely fast hand-off, allows micro cells to be used, and allows a smooth hand-off at high vehicular speeds. The spread spectrum technology, provides complete privacy to your phone calls. No one will be able to listen in, and understand your phone message, certainly not in real time.

The value to a service provider, is that broadband CDMA permits the largest number of users per square mile, and provides a lowest cost for handset and base. Indeed, we have spoken to chip manufacturers in the United States and they estimate that a handset using Broadband spread spectrum CDMA would cost less than $100 for the handset. A comparable handset using TDMA would cost close to $1,000.

Worldwide Advances in Communication Networks
Edited by B. Jabbari, Plenum Press, New York, 1994

85

PARAMETER	FDMA	TDMA	GSM	B-CDMA
Bandwidth (MHz)	12.5	25	25	48
Frequency Reuse	7	4		1
Users/Cell	56	1250	110	2000
Spectral Efficiency (Erlangs/MHz)	4.5	50		80
Density (Erlangs/km^2)	2	480		30,000
Area Efficiency (Erlangs/km^2 - MHz)	0.2	19	4.4	600

*Courtesy Westinghouse Corporation

Figure 1. PCN TYPICAL DESIGN PARAMETERS.*

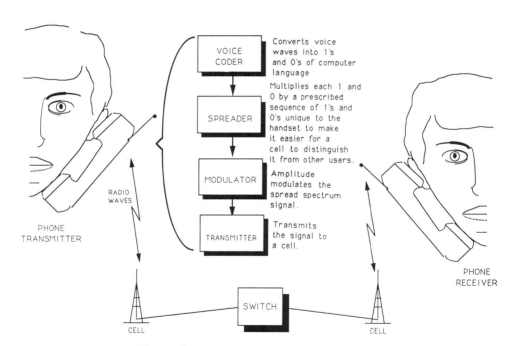

Figure 2. How Code Division Multiple Acess works.

Figure 1 compares the proposed FDMA, TDMA and Broadband CDMA systems being considered for use in the PCN band. Note that broadband CDMA is far more efficient than either of the other systems.

Figure 2 illustrates how Code Division multiple access operates. Consider that the phone transmitter contains a voice coder, which converts voice waves into 1's and 0's of computer language. Each 1 or 0 is multiplied by a high frequency sequence of 1's and 0's, that are set up in a unique, pseudo random, manner so that the cell can distinguish this particular user, from all others. This sequence is called the spreading sequence.

In the broadband CDMA PCN system, there are 750 spreading bits for every voice bit of data. We call the spreading bits, chips, since they "chop up" or <u>chip</u> the bit. We say there are 750 chips, per bit of data. These chips form a pseudo random sequence, and each sequence is almost orthogonal to the others. This results in orthogonality between

- Frequency Band : 1850 - 1990 MHz
- Chip Rate : 24 Mchips/s
- Bandwidth : 48 MHz
- Voice Coding : 32 kb/s (uncoded), Adaptive Delta Modulation
- Encryption : optional
- Processing Gain : 750
- Handoff : Smooth, Initiated by the user
- Location : 40 Feet

Figure 3. B-CDMA [sm] Features and Specifications.

users. The output of the spreader, is then modulated using binary phase shift keying, and transmitted to the cell.

In this system the cells are connected together to the switch and the receiver demodulates, despreads, and decodes, in order to get the voice back again.

The features of this spread spectrum system are outlined in Figure 3. The spreading code rate, or chip rate is 24 Mchips/s. Because binary phase shift keying is used, the bandwidth is 48 MHz, although narrower bandwidths could be used. Voice coding is 32 kb/s, and Adaptive Delta Modulation, with no forward error correction coding, is used. Encryption is optional and is really not needed. The processing gain is 750 chips per bit of data, or 29 dB.

Hand-off is fast and smooth and initiated by the user. Furthermore, a party can be localized by the base station to within 40 feet and using sectorized antennas, to 90 degrees.

Figure 4 shows the spectrum of the broadband CDMA transmitter. There is no multipath in this system, and the side lobes have been removed by filtering. The center

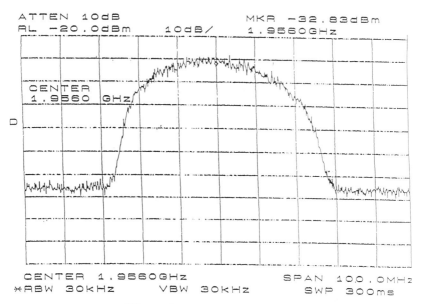

Figure 4. Spectrum Analyzer Plot.

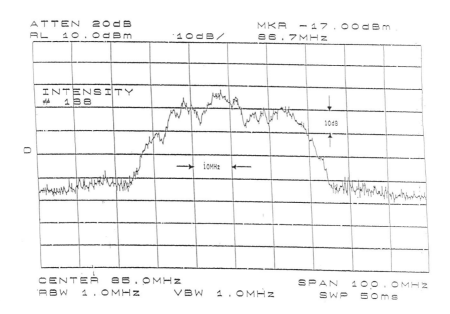

SPECTRUM ANALYZER PLOT, SITE #6, RECEIVER RF MONITOR

COURTYARD OF THE WTC

Figure 5. Spectrum Analyzer Plot with multipath.

frequency is 1.956 Ghz, the frequency deviation is 10 MHz per division, and 10 dB attenuation per division.

Figure 5 shows what happens when there is multipath. This picture was taken outdoors in the courtyard of the World Trade Center, that is between the two, twin-tower buildings. The multipath received can be as wide as 10 MHz. Broadband CDMA operates properly in this multipath environment, while TDMA, FDMA, and narrowband CDMA systems fail to operate. They need fade margins of 30 dB to 40 dB, while this system has a fade margin of about 1 to 2 dB.

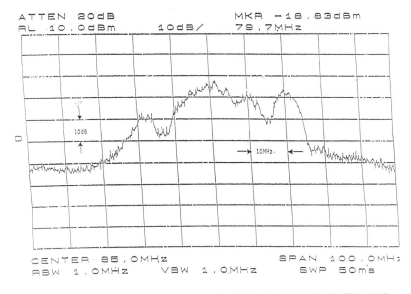

SPECTRUM ANALYZER PLOT, HANDSET #2, SITE #3, RECEIVER MONITOR PORT

INDOOR IN THE OFFICES OF A MAJOR INVESTMENT BANKER

Figure 6. Spectrum Analyzer Plot taken inside the offices of a Major Investment Banker in New York.

Figure 6 reinforces this result, by looking at the spectrum taken inside the offices of a major investment banker in New York City. Note that much broader fades occur. Once again, the total amount of fading of this picture, compared to the transmitted wave form shown in Figure 4 is less than 2 dB. Indeed, if a whole sideband was removed by fading, 25 MHz, there would still be single sideband transmission. That would only be a 3 dB fade.

CONCLUSIONS

In conclusion, PCN will be competitive to existing cellular and wired systems. It will provide high quality, low outage times, and toll quality voice. It will allow high data rate transmission. And, it will increase the number of users, significantly.

CONTROL ARCHITECTURES FOR FIXED AND MOBILE NETWORKS

Bijan Jabbari and Sirin Tekinay

Department of Electrical and Computer Engineering
George Mason University
Fairfax, VA 22030

ABSTRACT

Providing advanced network services to subscribers in fixed and mobile cellular networks requires enhanced call control and network features as well as mobility management. The support of these requirements necessitates a solution based on the Intelligent Network concept. For example, in cellular and microcellular networks as well as in fixed networks, subscriber mobility may require multiple database lookups and additional facilities to locate, register and authenticate users. The signalling load and the real time performance may be significantly hampered without the proper architecture. In this paper, we will address the important issues in network control assuming an architectural framework based on the intelligent network. We will consider representative control functions of mobile networks within the framework of this architecture and discuss control traffic behavior and performance.

1. INTRODUCTION

Network control plays an important role in both fixed and mobile networks. In fixed networks, whether narrowband or broadband, routing of calls, congestion control, handling of multimedia calls, enhanced service introduction and modification are of significant concern. In mobile networks, call and mobility control to provide roaming, handover, registration, and traffic and channel management, especially in the presence of heterogeneous classes of users and diverse access methods are major network control problems to be addressed. Somewhat related to both types of networks is the network intelligence distribution and traffic loading due to call and mobility control.

Traditional fixed and mobile networks have limitations in their network control. In both types of networks, control has been based on an ad hoc approach. The main reason for this has been the significant reliance on the intelligence residing on the switch and building additional capabilities on top of the switching system. This in turn has been the result of a lack of standard open interfaces and lack of a common nonmonolithic architecture to allow the required flexibility. Some of these problems may still exist as we migrate towards integrated services digital networks (ISDN) and broadband ISDN.

Today's mobile networks mainly deal with providing telephony services in limited geographic service areas. These networks have limited roaming capability (for some carriers in presubscribed service areas) and are unable to provide seamless roaming. Some networks treat handover requests and the arriving calls equally, hence may cause dropping of ongoing calls when the network is heavily loaded and no resources are available. Some lack power control for radio channels. Growth within these networks may result in significant control traffic, in particular as the cell size is reduced (for example, introduction of microcells), or as the traffic pattern changes. Therefore, traffic handling in these networks will become quite complex, performance will suffer and the capabilities will be curtailed. As we are moving towards implementing the second generation of mobile networks (typified by digital systems), these limitations are being considered more seriously.

In this paper, after discussing the general problems present in network control and the limitations of traditional fixed and mobile networks, we focus on common goals of fixed and mobile networks mainly from control and traffic aspects. Subsequently, we will discuss a few important aspects of control architecture for mobile networks as well as the control traffic.

2. COMMON GOALS OF FIXED AND MOBILE NETWORKS

Fixed and mobile networks have a number of common goals in providing call and mobility control. Here, we consider fixed networks potentially capable of providing subscriber mobility, i.e., the required capability for networking regardless of the location of the subscriber whereas mobile networks have the capability to support terminal mobility. We discuss several representative commonalties.

First, both networks allow some degree of freedom in service usage. Fixed networks are to provide terminal independent services, including incoming call management and outgoing call management. Mobile networks provide the required capability for networking regardless of the location of the mobile terminal in any service area, at any given time.

Second, both networks require sophisticated call control and routing. This is for homogeneous as well as heterogeneous traffic classes. However, the latter could present a challenge. Internetworking is also a major goal for both networks and the control architecture extends to include this.

Third, both networks have priorities in service features although the priorities in fixed networks may differ from those in mobile networks. This priority in service features provides a good argument for flexibility in the control architecture.

The above common goals can be further dealt with at a detailed level. However, we only describe the major attributes of a control architecture and limit our discussion to a representative case in mobile networks to see how performance is affected through a particular set of capabilities.

3. ATTRIBUTES OF THE CONTROL ARCHITECTURE

A particular framework which has shown great promise in providing the flexibility needed in the control function as well as the services in fixed and mobile networks is the concept of intelligent networks [1-6]. This concept takes into account the Signalling System 7 (SS7) for transport of control information and database look-ups in connection with mobility management. Functions include location identifications, authentication, routing, call origination, call completion, call handover, roaming, and charging. These functions affect the performance, traffic handling, system efficiency and network capability.

There are a number of advantages with this architecture: First, it provides opportunities for effective routing and call handling (e.g., roaming and handover). Second it can address the problem of subscriber mobility (in addition to terminal mobility). Last, it facilitates control of heterogeneous traffic. In the following we consider sample control functions which can be accommodated within the architectural framework above. The emphasis of these functions will be in mobile networks.

4. EXAMPLES OF CALL AND MOBILITY CONTROL

In mobile networks the problem of keeping track of a subscriber as he moves throughout the service areas is an important problem. Through the registration process, we can locate and page a called mobile station. A mobile station is allowed to make and receive calls within or outside (roaming) the service areas. The network is responsible for routing incoming calls to the current service area and allowing the subscriber to make outgoing calls. Furthermore, as the mobile station moves across the cell boundaries, the calls in progress need to be maintained through handover to other cells.

Mobility management involves location registration, updating and other network functions to locate, identify and provide or continue service while the mobile station moves around (see Figure 1). Registration and handover each involve tradeoffs. Registration requires considering control traffic loading and network intelligence distribution. Handover requires considering the quality of service and traffic channel loading and management, respectively. The control traffic relates to location updating and paging messages and involves signalling over radio channels as well as between MSCs (Mobile Switching Center) and network databases such as HLRs (Home Location Register) and VLRs (Visiting Location Register). As an illustration of the messaging involved, a typical location registration process is described below [6-8]:

(i) A mobile station enters into a service area served by a new MSC.
(ii) The mobile station requests a location registration from the network.
(iii) The visited MSC transfers the location registration request to the VLR.
(iv) The VLR allocates to the mobile station memory area and a roaming number.
(v) The VLR accesses the HLR and obtains the subscription information of the MS.
(vi) The VLR stores the LAI (Location Area Identifier).
(vii) The subscriber information in the old VLR is deleted.

Roaming implies that both networks can access the user class information in the HLRs/VLRs of each other's networks. When a subscriber makes outgoing calls, the VLR uses the information on the subscriber to enforce the correct calling privileges. For receiving incoming calls however, the originating switch or the switch which anchors the subscriber's directory number needs to query the HLR for a routing number to which the incoming call can be routed. The switch whose coverage area the called party has roamed into, needs to assign a temporary number to receive the incoming call. It then needs to alert the subscriber and upon answer, connect him to the caller. A typical process of call completion to a roaming subscriber is described below[6,7]:

(i) A call is routed to the gateway MSC of the called mobile station's home network based on the dialed directory number (e.g., MSISDN- Mobile Subscriber ISDN Number).

(ii) The gateway MSC accesses the HLR and obtains the serving VLR.

(iii) The HLR requests routing information from the VLR.

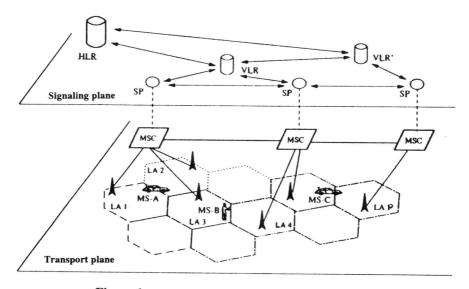

Figure 1. Mobility management problem in mobile networks

(iv) The VLR forwards this message to the visited MSC.

(v) The visited MSC sends routing information to the HLR.

(vi) The HLR forwards this information to the gateway MSC.

Handover strategies are categorized as follows (see Figure 2). In the nonprioritized scheme, no distinction is made between handover calls and newly arriving calls. Therefore, a call with a need for a channel in the crossed cell will be forced to be dropped if there are no channels available. This may have a significant adverse effect on the perceived quality of service. Thus, guard channels are introduced to improve the probability of forced termination at the expense of some inefficiency in spectrum utilization. Another category of handover is queueing of the handover requests which overcomes the problems associated with the spectrum inefficiency. To further improve the probability of forced termination, dynamic priority queueing may be utilized (see Figure 3) where the available channel is assigned to handover requests with high instantaneous priority.

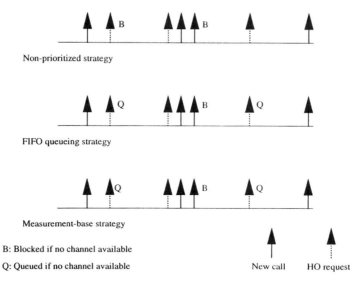

Figure 2. Handover strategies

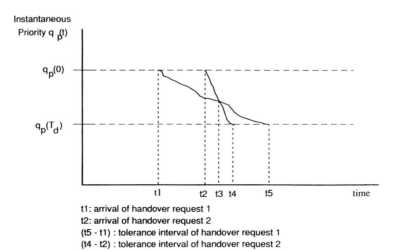

t1: arrival of handover request 1
t2: arrival of handover request 2
(t5 - t1) : tolerance interval of handover request 1
(t4 - t2) : tolerance interval of handover request 2

Figure 3. Dynamic priority -based handover

5. PERFORMANCE ASPECT

In this section we discuss the network performance aspect due to the representative control functions discussed above. The important performance parameters related to handover are: probability of call blocking, probability of handover failure and probability of forced termination. In Figures 4 and 5, we present the probability of call blocking and forced termination for three cases of non-prioritized, FIFO, and dynamic priority (designated as Measurement-Based Queueing in the Figure) for a range of offered traffic in each cell when handover traffic is 20% [10]. The Figures show the advantage gained clearly through dynamic priority. However, one needs to determine the handover traffic rate or the probability of handover in a given mobile network taking into account the mobility of users. Therefore, models are needed to capture the mobility pattern of the users. Such a mobility model not only can give us an estimate of probability of handover, but also can yield other control related functions such as paging, location updates and ultimately help us in evaluation of the system capacity and other related network traffic parameters such as channel holding time.

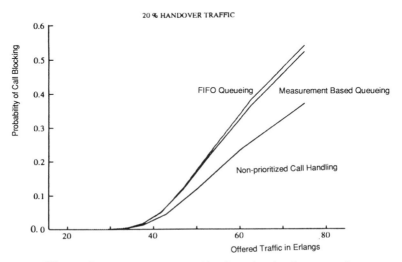

Figure 4. Probability of call blocking for various handover strategies

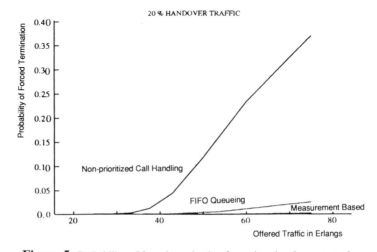

Figure 5. Probability of forced termination for various handover strategies

As an example, in this section we discuss the computation of the transaction intensity imposed on the information handling system as typically carried out in a simplified design [11]. Define transaction intensity as the number of demands raised per unit time which call for a certain number of transactions to be performed. The number of required transactions for each demand depends on the specific network architecture. Possible network architectures range from hierarchical to distributed information handling systems. A hierarchical configuration would be represented by a central dynamic control unit serving as both the HLR and the VLR for several MSCs. The distributed information handling strategy, on the other hand, can be depicted by MSCs performing the functions of the HLR and the VLR for the location area(s) they serve.

We consider the transaction intensity due to location updates. While the number of transactions carried out per location update is peculiar to the configuration of the information handling system, the intensity of location updates is a function of the number and average velocity of mobiles, cell sizes and the number of cells per location area.

Assuming that mobiles move with constant velocity where direction is drawn from a uniform distribution between 0 and 2π, the rate h of location area boundary crossings is given by :

$$h = \frac{n \cdot v \cdot P}{\pi \cdot s}$$

where n is the total number of active mobiles, P is the perimeter of the area, v is the magnitude of velocity, and s is the surface area. The relationship is plotted in Figure 6 for a cell radius of 1 to 3 km as the number of cells per location area changes.

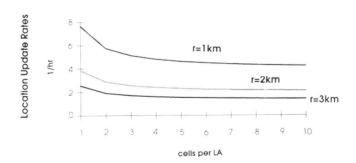

Figure 6. Location update rates for r = 1, 2 and 3 km

The simple mobility model above provides only loose estimates of cell or location area crossings based on movement of terminals in a uniformly distributed direction. In this case, the criterion of boundary crossing is the distance. In general, this criterion can be defined as any meaningful system parameter such as a signal power, signal to interference ratio, probability of bit error and associated specified thresholds or ranges for boundary crossing. In this situation, the boundary crossing is interpreted as threshold crossing for the criterion of choice. Therefore, more general mobility models are needed to capture the effect of the parameter and estimate the probability of boundary crossing. Such models are in particular needed for a system based on the microcellular concept where the smaller cell size may give rise to increased values, thus requiring a better estimate of boundary crossing. Moreover, a more accurate modeling of terminal movement may provide more meaningful estimates of probability of handover and location area crossing for when the criterion of choice is distance. One possible model is to consider the terminal movements within a cell as a random walk process. Based on this model, we can derive the distribution of the channel occupancy time, as well as the probability of boundary crossings and subsequently obtain the location area update rate.

6. CONCLUSION

The control architecture is a vital part of design aspect of future fixed and mobile networks. In this paper we have discussed the attributes of a control architecture framework based on the concept of intelligent networks and have discussed representative functions and their impact on network performance. Such a framework appears to provide effectively the essential functionality for future networks, namely mobility. Much research work is needed to quantify the benefits, evaluate the detailed control architecture alternatives and network intelligence distribution, and system traffic handling capacity. The mobility model discussed here can provide one approach in this research arena.

REFERENCES

[1] K. Murakami, M. Katoh, "Control Architecture for Next-Generation Communication Networks Based on Distributed Databases", *IEEE Journal on Selected Areas in Communications*, Vol. 7, No. 3, April 1989.

[2] M. Wizgall, W. Weiss, W. Stier, "The ISDN Approach for Mobile Radio", *Proceedings of the XIII International Switching Symposium*, 1990.

[3] M. Grenzhauser, H. Auspurg, "The Digital Mobile System D900 - A Step Towards the Telecommunication Management and Intelligent Network Architecture," *Proceedings of the XIII International Switching Symposium*, 1990.

[4] A. Nakajima, M. Eguchi, T. Arita, H. Takeda, "Intelligent Digital Mobile Communications Network Architecture", *Proceedings of the XIII International Switching Symposium*, 1990.

[5] M. Fujioka, S. Sakai, H. Yagi, "Hierarchical and Distributed Information Handling for UPT", *IEEE Network Magazine*, Nov. 1990.

[6] B. Jabbari, "Intelligent Network Concepts in Mobile Communications," *IEEE Communications Magazine*, Vol. 30, Feb. 1992.

[7] EIA Interim Standard, EIA/IS-41, Cellular Radiotelecommunications Intersystem Operations:, Washington, D.C., Feb 1991

[8] J. I. Yu, "IS-41 for Mobility management," *Proceedings of the IEEE ICUPC '92*, Dallas, TX, Sept. 1992.

[9] ETSI TC GSM, Recommendation GSM 03.12, Location Registration Procedures, Version 3.1.4, June 1990

[10] S. Tekinay, B. Jabbari, "Analysis of Measurement Based Prioritization Scheme for Handovers in Cellular Networks," Proceedings of the IEEE Globecom '92, Orlando, FL, Dec 1992.

[11] G. Morales-Andres, An approach to Modelling Subscriber Mobility in Cellular Radio Networks," Telecom Forum 87, Geneva, Nov. 1987.

MOBILE-USER NETWORKING: THE SATELLITE ALTERNATIVE

Anthony Ephremides

Electrical Engineering Dept.
University of Maryland
College Park, MD 20742

Jeffrey E. Wieselthier

Information Technology Division
Naval Research Laboratory
Washington, D.C. 20375

INTRODUCTION

It is clear that mobile or portable terminals must rely on wireless channel transmissions over communication media such as the satellite channel. In recent years, ambitious satellite systems have been under development that use "intelligent" space segments with on-board processing capabilities and that go beyond the traditional usage of satellites as "bent pipes." Network control schemes designed for such systems must be compatible with the constraints of satellite systems. Particularly important is the "latency" disadvantage of satellite transmission (i.e., the 0.27 sec. round-trip delay for geostationary orbits, although this can be significantly ameliorated in low-earth orbits).

Thus satellites offer viable alternatives for mobile users. The main challenges in developing practical systems lie in the "seamless" connectivity service that must be provided in hybrid environments that mix terrestrial and satellite components.

In this paper we consider a problem of service integration (i.e., voice and data) in a simple satellite system. We adopt the principle of "boundary" schemes (movable or fixed) as proposed for terrestrial systems and superpose it on the idea of interleaved-frame reservation-based channel access for satellite usage that was proposed in [1]. The interleaved-frame protocols are uniquely suited to satellite environments because they take advantage of the latency characteristics of satellite channels. Simple modifications of the basic version of such protocols permit the integration of voice and data by means of the boundary approach.

Integration of voice and data on any network aims at striking a balance between the need to reduce voice-call blocking probability and the requirement for low data-packet delay. We consider a control mechanism under which the decision on whether or not to accept a voice call depends on the number of calls in progress and the number of data packets backlogged in queue. It was shown in [2] that the minimization of a weighted cost function of the form $E[D]+\alpha P_B$ (where $E[D]$ represents average packet delay, P_B

stands for blocking probability, and α is a fixed constant) yields a simple "switching-curve" admission control strategy; i.e., the state space is divided into "accept" and "reject" regions by a simple staircase-like contour that can be computed. The boundary scheme represents a gross approximation to such an "optimal" admission control strategy; however, it is easily implementable.

In this paper we describe the operation of the Non-Interleaved-Frame Fixed-Length (NIFFL) protocols, which have been developed for data-only operation. We show how they are easily extended to integrated voice/data operation, and we outline the development of the Markov model for the integrated protocol. Finally, we present performance results. Both P_B and $E[D]$ can be computed exactly for the framed protocols we propose. Furthermore, there is an optimal value for the boundary threshold that corresponds to a suboptimally minimum value of the weighted cost criterion discussed above.

NIFFL PROTOCOLS FOR DATA-ONLY OPERATION

Before describing the integrated protocols, it is useful to consider the communication system for data-only operation. It is then straightforward to extend the model to integrated networks. We consider M ground-based users (terminals) that communicate among themselves via a transponder, which broadcasts all messages it receives to all members of the user population. Data traffic consists of fixed-length packets, each of which requires one time "slot" for transmission. We define R to be the round-trip propagation delay, measured in terms of slot duration. It is assumed that each user has an infinite buffer in which it stores the arriving packets, which are assumed to form a Bernoulli process with rate λ in every slot. The total arrival rate is, therefore, $M\lambda$ packets per slot, which is equal to the throughput rate under stable operation since no packets are rejected.

The NIFFL protocols are modifications of the Interleaved-Frame Flush-Out (IFFO) protocols [1]. Under the IFFO schemes, the frame length varies in response to data traffic, thus permitting efficient operation at high throughput rates. The NIFFL schemes are based on the use of a fixed frame length, which maintains the constant delay that is desirable in voice systems. The IFFO and NIFFL protocols are based on a reservation structure in which the unreserved slots may be used for transmission on a contention basis. In this paper we limit our discussion to the pure-reservation version of the NIFFL protocol (PR-NIFFL) in which unreserved slots remain idle. The frame length is kept fixed at L slots, where $L \geq R$. The first slot of each frame, which consists of M "minislots" that are exclusively allocated to the M terminals in (contention-free) TDMA fashion, is known as the "status" slot; it is used by each of the terminals to reserve a transmission slot for each of the packets that were generated in the previous frame. It is assumed that all reservation minipackets are received successfully by all terminals following the round-trip propagation delay of R slots. We define:

R_k = total number of reservations in queue at beginning of frame k;
A_k = total number of packet arrivals in frame k, summed over all terminals.

Since each frame has a length of exactly L slots, the reservation information generated at the beginning of the frame is received before the end of the first slot of the next frame. If

$R_k > L - 1$, the $\tilde{R}_k = \max \ [(R_k - L + 1) \ , \ 0]$ "excess" packets are postponed until frame $k+1$.

The operation of the NIFFL protocols is illustrated in Fig. 1. Reservation minipackets for all packets that arrive in frame k are transmitted in the first slot of frame $k+1$. The quantity of interest that needs to be tracked is R_k, which evolves as a first-order Markov chain under PR-NIFFL. The number of reservations in the system at the beginning of frame $k+2$ is $R_{k+2} = A_k + \tilde{R}_{k+1}$. A complete discussion of the dynamics and the resulting transition probabilities is presented in [3].

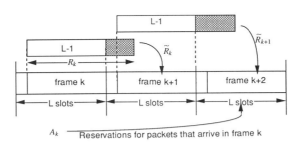

Figure 1. Operation of NIFFL.

VD-NIFFL PROTOCOLS FOR INTEGRATED VOICE/DATA SYSTEMS

We assume that voice calls are generated by idle users according to a Bernoulli process, and that they are geometrically distributed in length; thus the probability that a call is completed in any particular frame is also a Bernoulli process. The time constants associated with voice traffic are considerably larger than those associated with data traffic; voice calls will typically last from tens to hundreds of frames.

To accommodate the needs of both voice and data traffic, we consider a channel-access protocol under which a reservation scheme is used for voice traffic and PR-NIFFL is used for data. We call this the Pure-Reservation Voice/Data NIFFL (PR-VD-NIFFL) protocol (a similar scheme was studied in [4]). Under this scheme, once a voice call is accepted, it is guaranteed access to one slot in each frame until its completion. A fixed-length frame structure is necessary to accommodate the real-time requirements of voice traffic, and is appropriate for both satellite and ground-radio environments. The standard idea of a "boundary" mechanism is used to partition each frame between voice and data operation, as shown in Fig. 2.

Voice calls are accepted by the system as long as the total number of calls in progress simultaneously does not exceed some specified value V_{max}. If a slot is not available for a new call, the call is lost; there is no buffering of voice calls. Under the movable-boundary implementation, the slots not used by voice calls (including empty slots in the voice portion of the frame) are used for data traffic, which is transmitted by using one of the NIFFL protocols. Since the decision to accept new voice calls depends only on whether or not the threshold V_{max} is exceeded, voice traffic is unaffected by data traffic; however, the operation of the data protocol is dependent on voice traffic because data traffic is permitted to use unneeded voice slots.

Under a fixed-boundary implementation of the VD-NIFFL protocols, a fixed number of slots are available for data traffic in every frame. In this case the data-packet process is independent of the voice-call process, and the overall integrated system functions as two totally independent subsystems, one for voice and the other for data. Our primary interest is in the movable-boundary implementation because of its ability to provide improved performance and because its analysis is considerably more challenging.

As shown in Fig. 2, the first slot of every frame is once again the status slot, during which each terminal transmits its reservations for packets that arrived in the previous frame. Under a movable-boundary implementation, the next V_k slots are reserved for voice traffic, where V_k is the number of voice calls in progress at the beginning of slot k. The remainder of the frame consists of D_k data slots, where $D_k = L - 1 - V_k$. (Under a fixed-boundary implementation, $D_k = L - 1 - V_{max}$, independent of V_k.) As with the

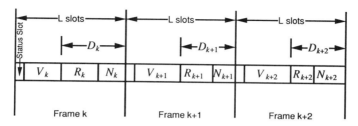

Figure 2. Frame Structure for the VD-NIFFL protocols.

protocols designed purely for data, R_k is the number of data packets for which reservations are needed at the beginning of frame k. Whenever $R_k > D_k$, the excess packets are delayed until frame $k+1$. Operation of the data portion of VD-NIFFL can thus be viewed as that of NIFFL with a variable number of slots (D_k) available for data traffic, where D_k depends on V_k.

AN EXACT MARKOV CHAIN MODEL FOR VD-NIFFL

The VD-NIFFL protocol can be characterized by the Markov chain (R_k, V_k), which has transition probabilities $Pr(R_{k+1}, V_{k+1}, | R_k, V_k)$. The development of a Markov chain model for the VD-NIFFL protocols has taken into account the dependence of data traffic on voice, whereas voice is independent of data. Thus

$$Pr(R_{k+1}, V_{k+1} | R_k, V_k) = Pr(R_{k+1} | R_k, V_k) Pr(V_{k+1} | V_k).$$

The transition from R_k to R_{k+1} depends on V_k in a movable-boundary system (because V_k determines D_k), but not on V_{k+1}.[1] The transition from V_k to V_{k+1} does not depend on R_k or R_{k+1}. Consequently, the transitions corresponding to the data process can be considered separately for each value of V_k. Thus it is not necessary to work with the huge transition probability matrix that characterizes the evolution of the complete voice/data state description.

We first consider the data transitions. Corresponding to each value of V_k there is an $N \times N$ transition probability matrix for the data message process with elements

$$p_{ij}^{v} = Pr(R_{k+1} = j \mid R_k = i, V_k = v).$$

Next, we consider the voice transitions. The probability that a new call is generated at a terminal during any particular frame is denoted as λ_V. We can view the M_V voice terminals as concentrators, each of which can support several (up to V_{max}) voice calls simultaneously. Each terminal is able to generate one new voice call in any given frame, with probability λ_V, independently of the number of voice calls it is already supporting. The probability that an ongoing call completes service during any particular frame is denoted as μ_V. Under these conditions the voice-call process may be characterized by a first-order Markov chain. Expressions for the transition probabilities, which incorporate both voice and data processes, are provided in [3] along with a discussion of how the equilibrium state distribution is obtained.

PERFORMANCE EVALUATION

The expected delay (or "system time") $E[D]$ of a packet is defined to be the expected time elapsed between the time a packet arrives at a terminal until it is successfully delivered to its destination. It depends not only on the number of other packets that are generated during the same frame, but also on the system state (R_k, V_k) at the time of its arrival. This is because the current backlog of data packets (R_k) must be transmitted before the new arrivals can be transmitted, and because the statistics of the number of data slots available in future frames (under the movable-boundary version) depend on V_k. Our approach, for both fixed- and movable-boundary schemes, is to determine the conditional expected system time, given the state (R_k, V_k), and then average this over the probability distribution of the state. Expressions for conditional expected system time are given in [3] for both fixed- and movable-boundary models.

The performance measure used for voice traffic is the voice-call blocking probability P_B. Whenever a slot is not available for a new call, the call is blocked and dropped from the system, independently of the data-packet backlog. Because the voice process, and hence P_B, is independent of the data-traffic parameters, P_B is the same for fixed- and movable-boundary versions of the protocol. A complete derivation of P_B is presented in [3].

Both data-packet delay and voice-call blocking probability have been evaluated computationally. All of the results presented here are for $M = M_V = 10$ and $L = R = 12$. Our earlier studies of the IFFO protocols [1] have shown that performance is quite insensitive to the number of terminals for M greater than about 5, and this behavior has also been observed for the PR-VD-NIFFL protocol; thus the results for $M = 10$ are representative of higher values as well.

---

[1]In a fixed boundary system, data transitions are independant of the voice process.

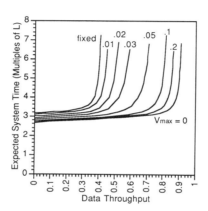

Figure 3. Normalized expected system time for data packets under movable-boundary PR-VD-NIFFL as a function of data throughput, showing dependence on μ_V ($L = 12$, $M = 10$, $M_V = 10$, $V_{max} = 6$, $\lambda_V = 0.01$).

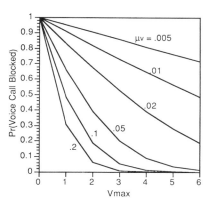

Figure 4. Voice-call blocking probability under PR-VD-NIFFL as a function of V_{max} ($M_V = 10$, $\lambda_V = 0.01$).

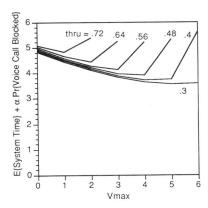

Figure 5. Weighted performance index for fixed-boundary PR-VD-NIFFL for $\alpha = 2$ ($L = 12$, $M = 10$, $M_V = 10$).

Figure 3 shows $E[D]$ (expressed in multiples of L) as a function of data throughput. The curves show the dependence of $E[D]$ on the voice process by varying μ_V over a wide range for the movable-boundary scheme with $V_{max} = 6$, while λ_V is kept fixed at 0.01. For small values of μ_V, the system is heavily loaded, and performance approaches that of the fixed-boundary scheme. As μ_V increases, the average length of voice calls decreases, resulting in a decrease in the voice-call load and hence an increase in the number of slots available for data traffic. Thus the system is able to support higher levels of data traffic. The rightmost curve, i.e., that for $V_{max} = 0$, represents the limiting case in which the entire channel resource is available for data traffic.

Figure 4 shows P_B as a function of V_{max} for $M_V = 10$, $\lambda_V = 0.01$, and several values of μ_V. Clearly, increasing V_{max} results in lower blocking probability because more of the channel resource is available for the voice calls. Also, since large values of μ_V correspond to short voice calls, increasing μ_V results in lower channel utilization and hence in lower blocking probability.

The minimization of data-packet delay and the minimization of voice-call blocking probability are conflicting goals. We have considered the weighted sum $E[D]+\alpha P_B$, where the coefficient α is chosen to reflect the relative importance of blocking probability and packet delay. Figure 5 shows this weighted performance index as a function of V_{max} for the fixed-boundary version with $\alpha = 2$, where delay is again normalized with respect to L. Each curve represents a fixed value of data-packet throughput. Note that for throughput values of 0.48 and greater, the curves terminate at the highest value of V_{max} (< 6) that will support the throughput. For $\alpha = 0$, only the expected system time is of concern, and the best performance is obtained for $V_{max} = 0$. For $\alpha = 2$, the trade-offs between the two performance indexes are apparent; as throughput decreases, the optimal value of V_{max} increases. As α increases, more importance is assigned to P_B, and higher values of V_{max} are appropriate for some of the higher values of throughput. For significantly higher values of α, it is best to use the highest value of V_{max} that will support the required data-packet throughput.

REFERENCES

[1] Wieselthier, J. E. and A. Ephremides, "A New Class of Protocols for Multiple Access in Satellite Networks," IEEE Transactions on Automatic Control, AC-25 pp.865-879, 1980.
[2] I. Viniotis and A. Ephremides, "On the Optimal Dynamic Switching of Voice and Data in Communication Networks," IEEE Computer Networking Symposium, pp. 8-16, April 1988.
[3] Wieselthier, J. E. and A. Ephremides, "A Study of Channel-Access Schemes for Integrated Voice/Data Radio Networks," NRL Report 9359, Naval Research Laboratory, Washington D.C., November 1991.
[4] Wieselthier, J. E. and A. Ephremides, "A Movable-Boundary Channel-Access Scheme for Integrated Voice/Data Networks," Proceedings of IEEE INFOCOM'91, pp. 721-731, 1991.

ADVANCES IN SATELLITE NETWORKS

Dr. W. W. Wu

The Consultare Group, Inc.
Bethesda, MD 20814

ABSTRACT

This paper addresses the issues of satellite communications related to networks. The paper brings forth some present problems, potential solutions, and future challenges. Specifically, the paper clarifies the myth of satellite over the hills, identifies the tradeoffs of onboard processing and switching, provides alternatives in lower earth orbit satellite networks, and highlights the principal architecture of the worldwide operational digital high speed satellite network. The paper also emphasizes the importance of integrated services digital satellite networks.

INTRODUCTION

Satellite communication is a multifaceted subject that encompasses many diversified and specialized practical disciplines. Due to its impact on large coverage and population, various application and usefulness, satellite communication is no longer in the domain of the technologists alone. It has been recognized that other factors, such as economical, institutional, policy, and regulatory, affect the advancement, development and usefulness of satellite communication. These issues, which no longer can be ignored, affect not only the application aspect, but also in research and development. These non technical factors, in many instances, delay the tasks or derail the directions on what should be pursued. In this paper the following questions are discussed:

- Are satellites over the hills?

- What are the network problems unique to satellites?

- What are the potential solutions?

- What are the present states of advancement?

- What are the future directions?

Section 2 answers the question why satellite communication is not over the hills. Section 3 discusses some problems which are unique to satellite networks. Section 4 attempts to suggest a number of areas of solutions. Section 5 outlines three lower earth orbit satellite networks. Section 6 highlights the INTELSAT TDMA NETWORK. Section 7 mentions the importance of integrated services digital satellite network concept. Section 8 provides some concluding remarks. Section 9 lists relevant references.

ARE SATELLITES OVER THE HILLS?

The answer depends on who you ask. If you ask some bean counters, and uninformed investors, the answer is yes. On the other hand, if you ask the three NASA astronauts Richard Hieb, Thomas Akers and Pierre Thuot, the answer is a no. Because today (May 14, 1992) as this paper is presented, the astronauts are space walking for the last effort in order to recover the INTELSAT VI spacecraft which was misfired into the wrong orbit a year ago. The task is not only challenging, but also dangerous.

Ever since fiber optical cables became a reality, the myth of satellites becoming limited usage with mature technology has been overwhelming. It is true that any technology over 25 years old should be considered as old hats. As the environment and applications of satellite communications change from large stations to miniature ones, from stationary to mobile, from a few users to thousands in a network, the network strategy, the access technique, and the method of modulation have not changed much from the days of trunk routes. As the environment of satellite communication changes, the techniques have not changed accordingly. As a consequence, as if to kill a chicken using an ax for oxen, the incompatibility of over designed systems are not cost effective. In turn, the usefulness of these systems cannot be widely afforded.

Therefore by recognizing satellite communication in a changing environment, new applications require new methods of transmission, and alternatives of network architectures for the purpose of reliable low cost systems. Instead of over the hills, satellite communications has opened up with new technical demands with challenges. This is particularly true for network related disciplines, which has been ignored in the earlier development of satellite communication.

NETWORK PROBLEMS UNIQUE TO SATELLITES

Traditional satellite communication has no significant network problems. Because stations were few in number and large in size. Almost all satellite networks were centrally controlled, and most communications were point to point. Throughput was pre-arranged and queueing was not allowed. The propagation delay of geosynchronous altitude is a fact of life and no network strategy can improve upon. Most satellite networks such as INTELSAT and INMARSAT networks operate on system basis. Except in monitoring situations all systems are operated independently. Thus efficiency and flexibility of inter-system operations do not exist.

Network performance is usually measured by throughput and time delay. In addition to these two reference measures satellite networks have the following specific concerns:

To all earth stations, a satellite appears as a network node with limited throughput.

- The propagation delays cause incompatible protocols which have been recommended by the international standards committee.

- For free destinational short packet transmission the synchronization time is excessive.

- Combined interference problems.
  - Adjacent channel
  - Co-channel
  - Intersymbol

- Inter-satellite networking.

- Elevation angle and transmission speed limitations.

- Launch vehicle compatibility.

- Orbital Determination.

From network viewpoint only the selected few from above are discussed in the following:

Satellites as Network Bottlenecks

The subject of packet-satellite has been addressed by many [1-7]. The original idea of packet transmission for radio and terrestrial data link has enjoyed network flexibility, operational efficiency, and resource conservation through multiple path allocation for transmission of short length packets. However, these attributes have not been realized for satellite communications with packets. For satellites, even with multiple transponders and multiple onboard processors, have appeared as bottlenecks. In principle, the merit of multiple pathed short packet transmission is lost, when it applies to satellite communication. Thus, one of the challenging problems is the elimination of this bottleneck effect.

Propagation Delays

The transmission delay from an earth station to a satellite is a function of the speed of light, radius of the earth, altitude of the satellite, and satellite coverage angle. In comparison with fiber optical cables the propagation delays of satellite links are excessive. Worse, for a given satellite altitude and coverage angle there appear little technologies can do to minimize such delays. In addition, the time delay multiplies from the single path delay when multi-hop satellite links are considered. Despite a large amount of efforts and the success of echo control mechanism for voice circuits, the fundamental problem of propagation delay in satellite networks remains.

Protocol Incompatibility

The Study Groups of the CCITT recommended the 7 layer Open System Interconnection (OSI) as well as the Asynchronous Transfer Mode (ATM) for data communication networks. OSI defines the required protocols to ensure efficient and reliable interface and transmission. In OSI protocols the standards demand fast acknowledgement for high speed computer or optical communication, and the recommended response time in the protocols is beyond the feasibility of the propagation delay in a satellite link.

As it is standardized the ATM calls for minimal level of error control in the ATM Header and none in the Information Field, which carries messages. In fiber optical channel such recommendation may be sufficient, but it is inadequate in severely interference limited as well as fading channels of fixed satellite or mobile satellite links.

Interferences

As multiple accessed users increase, and satellite communications extend services for VSATs, mobile terminals and personal communications, the limiting factor in information transmission is no longer the space channel characterized by additive Guassian noise alone. Rather, depending on the application, a satellite channel can be dominated by interferences, fading, shadowing, or their combination effect.

When a transponder is shared by multiple carriers and the isolation among the carriers is not perfect, adjacent channel interference (ACI) occurs. Two factors contribute to ACI. The first one, referred to as truncation noise, is caused by the combination of the transmitting and receiving bandpass filters' inability to accommodating all the spectrum energy of the carriers. The second type of ACI, called convolution noise, produces baseband noise and is caused by convolving the spectra of two adjacent modulated radio frequency carriers after filtering.

Cochannel interference (CCI) can be caused by insufficient spatial signal isolation or polarization isolation, where interfering frequencies are in-band with the desired signals. The level of CCI depends on the spacing of satellites, antenna beam isolation, sidelobe effect, and the imperfect management of frequency reuse.

Intersymbol interference (ISI) is not unique to satellite networks. ISI differs from ACI and CCI by the fact that the interference is not caused by external sources. ISI can be caused by none-optimum signal design, sampling, or filtering. Most ISI in practice is caused by imperfect channel filtering as a result of high speed signaling.

Fading characteristics become important when a satellite is used for mobile, small and personal terminals. For these types of receivers the environment is different from traditional larger earth stations. As a finite state transmission channel, most fading characteristics in radio communications are known, but the effect of the combination of above interferences, fading, and Gaussian noise all together is not so clear, particularly if the number of interferences is not large and the law of large numbers cannot apply.

Inter-Satellite Networking

Inter-satellite links (ISL) have been investigated for INTELSAT, NASA, and ESA geosynchronous satellite programs. On the ground optical research has been performed for inter-satellite links. One of the major problems has been the mechanical pointing accuracy for either optical or microwave inter-satellite transmission. The other problem is lack of justification. With large amount of information transmission capacity among ISL, it will not be a significant benefit if transmission capacity cannot be enhanced between satellite and ground stations. At optical frequencies the signal attenuation between a satellite and an earth station discourages many advocates.

However, with the advent of lower earth orbit (LEO) satellites, inter-satellite networking is an essential aspect of the architecture. With the spacecraft and launch vehicle technology matured, the unproven and also the most challenging part in a LEO program is inter-satellite networking. The success of any LEO system will be heavily depended upon ISL.

POTENTIAL SOLUTIONS

Network Architectures

In a terrestrial network of n nodes to communicate point to point, the required number of links is n!/2(n-2)!. For a satellite network the same n nodes only require n links. But the point to point approach, which has dominated satellite communications for decades, has over-shadowed the real strength of the usefulness of a satellite - the capability of broadcast. Thus, point to multi-point network architecture should be co-existent with point to point in satellite communication.

Control stations or sharing a hub station are part of the present VSAT, mobile satellite network strategy. But these larger stations contribute to most of the cost of a network. Thus in order to provide very low cost networks, the costly control stations must be eliminated. For most satellite data communication, such hubless network architecture is feasible.

Onboard Processing and Switching

Functions which can be performed onboard a satellite are:

- Demodulation and remodulation

- Error decoding and recoding

- Transponder hopping

- Interconnection of antenna beams

- Information storage and retrieval

The advantages of onboard processing and switching can be identified as follows:

- Error rate reduction

- Efficiency improvement

- Capacity enhancement

- Network connectivity

- Frequency reuse

- Minimization of response time

- Optimization of traffic assignment

- Interference cancellation

- Linearization and equalization

- Data reduction, sorting, routing, and distribution

The disadvantages with onboard processing and switching are: complexity, weight, heat dissipation, additional power requirement, size, and reliability due to added components.

Lower Earth Orbit Satellite Networks

One way to reduce propagation delay from geosynchronous orbit is to reduce the distance from a satellite to earth surface. As early as the 1960s lower earth orbit (LEO) satellite network was proposed, but it was the IRIDIUM system proposed by Motorola in 1990 that gave the satellite community a gust of fresh air after more than 25 years of geosynchronous activities [8].

LEO satellite networks are potentially attractive to mobile, two-way paging, cellular radio, voice and data of personal communications. The networks can also provide accurate location determination as well as for large number of sparsely placed VSATs.

In general the difference of total carrier-to-noise ratio, $DIF(C/N)_t$, between any two orbits in terms of power, path loss, interference is shown in the next page. Where :

DIF (e.i.r.p) = Difference in earth station or satellite effective isotropically radiated power.
DIF [PL(u,f)] = Difference of path loss of up-link at frequency f.
DIF [PL(d,f)] = Difference of path loss of down link at frequency f'.
DIF $(C/N)_{u \, or \, d}$ = Difference of carrier-to-noise ratio for up-link or down link.
DIF $(C/I)_{u \, or \, d}$ = Difference carrier-to-interference ratio for up-link or down link.

$$DIF(\frac{C}{N})_u = DIF(e.i.r.p) + DIF[PL(u,f)]$$

$$DIF(\frac{C}{N})_d = DIF(e.i.r.p) + DIF[PL(u,f')]$$

$$DIF(\frac{C}{N})_t = \{[DIF(\frac{C}{N})_u]^{-1} + [DIF(\frac{C}{N})_d]^{-1} + [DIF(\frac{C}{I})_u]^{-1} + [DIF(\frac{C}{I})_d]^{-1}\}^{-1}$$

The advantages and disadvantages of LEO in comparison with GEO in general are listed in Figure 1. The advantage of small user antennas listed assumes the same amount of satellite power in both cases. The inter-satellite switching and other complexity problems may be considered as challenges rather than disadvantages.

The effect of LEO can be expressed in terms of the difference of total carrier-to-noise ratio with respect to elevation angle, i.e. the angle measured between earth station - satellite and the local horizontal. An Example of the variation, data provided by JPL, is shown in Figure 2. The plot is evaluated at altitude 5.1×10^6 m in comparison with GEO at 3.578×10^7 m. At 30 Ghz the path loss due to propagation is shown in Figure 3 for different altitudes. The decreasing amount of propagation loss due to lower altitude can be directly translated to increasing in power gain in the link budget.

ADVANTAGES OF LEO

- *SMALLER USER ANTENNAS*

- *REDUCE PROPAGATION DELAYS*

- *LESS RADIATION PROBLEM*

- *COST EFFECTIVE LAUNCH VEHICLES*

DISADVANTAGES OF LEO

- *INTER-SATELLITE SWITCHING*

- *HANDS-OVER COMPLEXITY*

- *NETWORK CONTROL PROBLEMS*

- *ATMOSPHERIC DRAG*

- *LESS COVERAGE*

Figure 1. Tradeoffs of LEO systems.

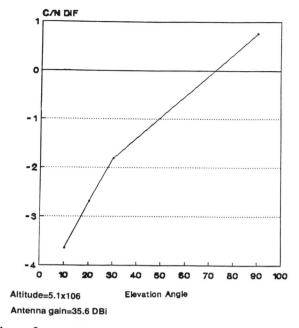

Figure 2. Total Difference Carrier to Noise Ratio vs. Elevation Angle.

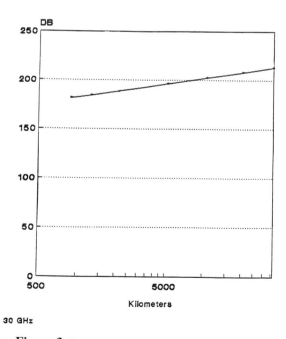

Figure 3. Propogation Loss vs. Distance in Space.

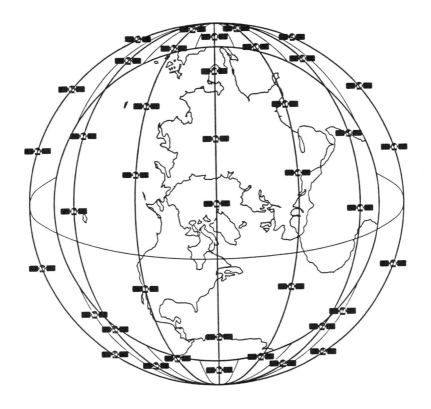

- POLAR ORBIT
- 765 Km ALTITUDE
- 77 SATELLITES

Figure 4. The IRIDIUM Constellation.

LEO SATELLITE SYSTEMS

Three LEO systems, the IRIDIUM, the GLOBALSTAR and the KRYPTON will be briefly mentioned. The coverage of IRIDIUM is shown in Figure 4, in which 7 circular orbits, each containing 11 satellites, will operate at the altitude of 765 km, or 413 nautical miles. The characteristics of IRIDIUM is outlined in the next page.

CHARACTERISTICS OF IRIDIUM

- GLOBAL COVERAGE, EQUALLY SPACED

- 7 CIRCULAR POLAR ORBITS

- 11 SATELLITES PER ORBITAL PLANE

- LOW EARTH ORBITS: 413 NM

- 14 SLICES ~ 26 DEGREE LATITUDE SPACING

- 33 DEGREE SATELLITE LONGITUDE SPACING

- ALL SATELLITES AMONG THE ORBITAL PLANES ROTATE UP TOWARD NORTH POLE AND DOWN TOWARD THE SOUTH POLE AT THE SPEED 7,400 M / S

- ORBITS 1,3,5,7 ARE IN PHASE 2,4,6

- EACH SATELLITE COVERS 37 CELLS (SPOT BEAMS)

- EACH CENTER CELL IS SURROUNDED BY 3 RINGS OF SMALLER CELLS

- THE 3 RINGS CONSIST OF 6, 12, 18 CELLS

- EACH CELL ~360 NM DIAMETER

- 77 X 37

- CELLULAR CONCEPT, DIFFERENCES

With 48 satellites in circular orbit the GLOBALSTAR is designed to be operated at an altitude of 1,398 km, or 750 nautical miles with global coverage. The 48 satellite constellation has 8 orbital planes with 6 satellites in each plane. Each orbital plane has an inclination angle of 52 degrees. The inclination angle is the angle between an orbital plane and the earth equatorial plane. With this constellation the average connection time for a user to a single satellite is about 10 minutes [9].

GLOBALSTAR is designed by Loral Aerospace to be operated in both C (5199-5216 Mhz for down link and 6525-6541 Mhz for up-link) band and L (1610-1626 Mhz) band. L-band is used for both user-to-satellite and satellite-to-user links, while C-band is used for both satellite-to-
gateway and gateway-to-satellite links. In this case both right-hand and left-hand circular polarization are used.

KRYPTON is a low cost LEO satellite network concept, which uses 36 satellites in a 63.43 degree inclined, elliptical orbit to cover the high density North Atlantic and North Pacific communication routes. The semi-major axis of 7,1439.9 km is the same as IRIDIUM for ease of launch. With 0.08 eccentricity, the apogee of KRYPTON is 1,337.3

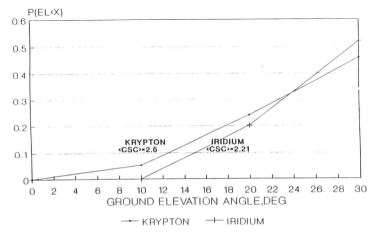

Figure 5. KRYPTON Elevation Statistics. (**30 to 70 Deg N. Latitude, Dense Traffic**)

km altitude and perigee is at 194.2 altitude. For 12 times during the 1.67 hour KRYPTON orbit, every spot in the viewing area is checked for the maximum elevation angle available [10].

By selecting slightly higher altitude than IRIDIUM the KRYPTON communication paths are out of the ionospheric scintillation region. In the meantime, less number of satellites will be used. Due to longer viewing time from the earth stations in the network, switching is less frequent. Because transoceanic regions are often within a orbital plane, satellite crosslinks are relatively simple to implement. The KRYPTON system design, which is different from either IRIDIUM or GLOBALSTAR, is described in [11].

Compare IRIDIUM and GLOBALSTAR, Figures 5 and 6 show the cumulative elevation angle statistics in the 30-70 degree North Latitude region. This is the region in which high air and sea as well as intercontinental communication traffic are intense. The curves in Figure 5 indicate that KRYPTION network stations need to use 10 degree elevation angle for 5% of the weighted trials. Which causes the ensemble average cosecant (CSC, elevation) of 2.5 for KRYPTON and 2.21 for IRIDIUM. This difference can provide a saving of half the space segment cost. Comparing with GLOBALSTAR Figure 6 shows identical probabilities at 10 and 24 degrees elevation angles. The coverages of both KRYPTON and GLOBALSTAR are excellent, but the KRYPTON network uses 48-36 = 12 satellites less than GLOBALSTAR.

THE INTELSAT TDMA NETWORK

The INTELSAT TDMA network consists of four reference stations, owned and

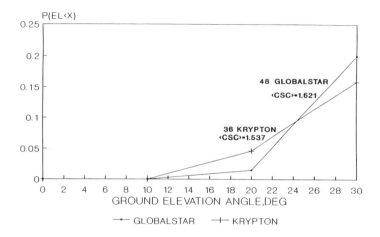

Figure 6. GLOBALSTAR Elevation Statistics. (**30 to 70 Deg N. to 70 Deg N. Latitude**)

operated by INTELSAT, two in each hemisphere covered by Zone beams. Up to 116 traffic stations in the network can be owned and operated by the participating countries. Presently in operation Figure 7 shows the network configuration. The figure contains the specific transponder assignment, traffic terminals (T) either in East Zone (EZ), West Zone (WZ), East Hemisphere (EH), or West Hemisphere (WH) beam. The satellite is located at 330.5 degree East. The locations of the stations are listed in Table 1. The traffic burst time plan is shown in Figure 8. Where scales are measured in kilo-symbols.

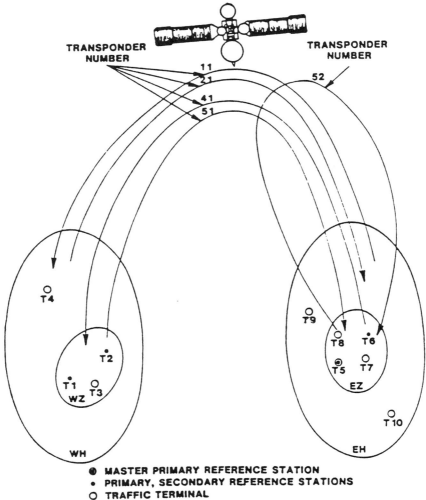

Figure 7. The Network Configuration.

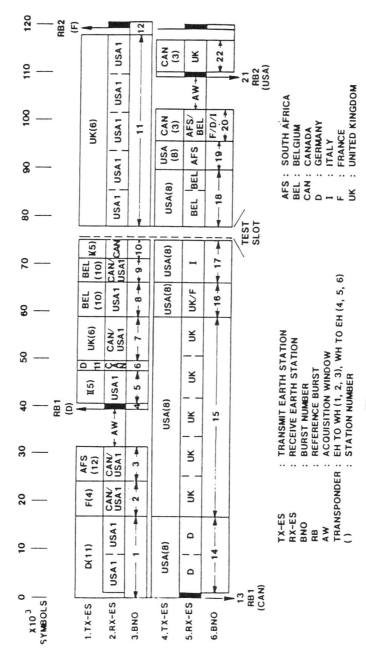

Figure 8. Traffic burst time plan.

Under INTELSAT sponsorship a TDMA Network Simulation (TNS) computer program was initiated by the author and developed jointly by KDD and Mitsubishi. As shown in Figure 9, the TNS program consists of Controllers, Models and Modules. The functions of each are as listed. The details are omitted here and some information can be found in [12].

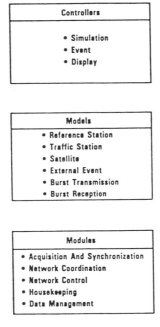

Figure 9. The Structure of INTELSAT TDMA Network Simulation Computer Program.

Integrated Services Digital Satellite Networks

ISDN is on the verge of reality. Whether it is the network provider, the equipment manufacturer, or the user, not complying with recommended international standards, regardless whether they are good or bad, will meet with failure. Communication by means of satellites is no exception. Thus future development in satellite network must take ISDN, whether narrow band or wide band, into consideration. As usage and complexity grow, a likely scenario is shown in Figure 10, where Signaling System Number 7 type structure will be part of the onboard switching and processing design, and all the ground terminals are transmitting and receiving with Asynchronous Transfer Mode format.

CONCLUDING REMARKS

Satellite communications has been developed mostly without the benefit of network principles. Which include optimization and low cost strategies. As applications of satellite communications change, network disciplines become more important for onboard processing, transponder switching, mobile and personal communications and for lower earth orbital systems.

On the other hand, through transponder traffic assignment, algorithms have been developed by using mini-max and graph theoretical methods. In order to minimize severe rain attenuation in a satellite channel, diversity by means of terrestrial networks, such as LAN, MAN, and WAN in connection with satellite links has been recently proposed.

Satellite communications will become more useful and economic, if network principles are considered and standardized protocols are applied. The good signs are that the bottom three layers of OSI have been implemented and the structure of ATM is investigated for digital satellite communications.

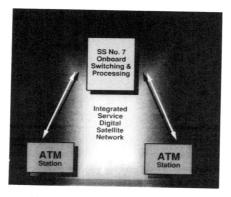

Figure 10. Onboard switching and processing design.

In terms of network activities the future directions can be:

• Almost all the themes, subjects discussed in this symposium can affect the advances of   satellite communication development.

• In some applications low cost hubless satellite networks will become a reality.

• Regional multiple small satellite networks may have stronger political and financial   support than global.

• Future operators of satellite networks and manufacturers of   satellite communication   equipment will comply the international standards for networking and protocols.

• Network interfaces between satellite and terrestrial will be unified for simplicity.

- Network optimization can be applied to:

    - Orbital locations for maximizing coverage and minimizing interference.

    - Minimizing the number of satellites for low cost network.

    - Inter-satellite links for connectivity and efficiency.

REFERENCES

[1] W. W. Wu, "Packetized Satellite Communication: Promises and Problems," Proceedings of the International Symposium on Signals, Systems, and Electronics, an invited paper presented at the Universitat Erlangen, Germany, September 18-20 1989.

[2] J. Chang, "A Multibeam Packet Satellite Using Random Access Technique," IEEE Transactions on Communications, Oct. 1983.

[3] T. Stevenson and K. W. Yates, "A New Multiple Access Scheme for Packet Satellite Communications," Proceedings of the International Symposium on Signals, Systems, and Electronics, Germany, September, 1989.

[4] H.Lee and J. Mark, "Combined Random/Reservation Access for Packet Switched Transmission Over a Satellite with Onboard Processing, "IEEE Trans. on Communications, Oct. 1984.

[5] N. Abramson, "Packet Switching with Satellites," National Computer Conference, 1973.

[6] I. Jacobs, E. V. Hoversten, and R. Binder, "General Purpose Packet Satellite Networks," Proc. of the IEEE, Nov. 1978.

[7] J. Derosa, L. Ozarow and Weiner, "Efficient Packet Satellite Communication," IEEE Trans. on Communications, Oct. 1979.

[8] R. J. Leopold, "Low Earth Orbit Global Cellular Communications Network," Proceedings of the Space Communications Technology Conference: Onboard Processing and Switching, NASA Lewis Research Center, Nov. 12-14, 1991.

[9] R. K. Kwan and P. A. Monte, "GLOBALSTAR: A New Mobil Communications System," Proceedings of the Space Communications Technology Conference: Onboard Processing and Switching, NASA Lewis Research Center, Nov.12-14, 1991.

[10] P. Christopher and W. W. Wu, "KRYPYON: A Low Cost Satellite Communication Concept," Proceedings of the National Telesystems Conference, May 19-20, 1992.

[11] W. W. Wu, and P. Christopher, "KRYPTON: Systems Designs," U. S. Patent Application.

[12] W. W. Wu, G. Forcina, K. Koga, H. Shinonaga, H. Mauro, G. Kondo, "The INTELSAT Network Simulation (TNS) Program," Proceedings of the International and Canadian Satellite Communications Conference, June, 1985.

[13] W. W. Wu, Guest Editor, Special Issue on ISDN, International Journal of Satellite Communications, Vol.9, No. 5, Oct. 1991.

SIGNAL CODING: THE NEXT DECADE

Nikil Jayant

Signal Processing Research Department
AT&T Bell Laboratories,
Murray Hill, New Jersey

ABSTRACT

The interworking of coding theory, signal processing and psycophysics has led to impressive capabilities in the compression of speech, audio, image and video signals. These advances have been accompanied by numerous standards for digital coding and communication, both national and international. Emerging technology goals require even greater levels of signal compression. As we address these goals, and as we seek to define and approach fundamental limits in coding, we are guided by several promising trends in compression research, some of which represent cross disciplinary evolution in the seemingly very different domains of acoustic and visual signals. Important aspects of new research in the field are the creation of refined methodologies for the measurement of signal distortion, and the definition of communications algorithms that make use of joint source and channel coding.

HIGHLIGHTS

Efficient compression of audiovisual signals is central to technologies for digital communication and information storage. A fairly comprehensive description of speech, audio, image and video applications is provided in Figure 1. In this illustration, the horizontal axis depicts bit rates after signal compression. The vertical axis has no special meaning. Projections of the application labels on the horizontal axis define typical bit rates or typical ranges of bit rates for levels of signal quality that are appropriate to the respective application. In entertainment applications such as music and television broadcast, demands on signal quality are generally much higher than in information applications. The bit rate capabilities in the figure represent high levels of compression in general. For example, an uncompressed audio signal is represented by 706 kbps on the compact-disc, compared to the 64 kbps capability shown in our illustration.

Recent advances in compression are due to a strong interworking of sophisticated coding algorithms and an exponential growth in arithmetic capability in hardware for digital signal processing (Figure 2).

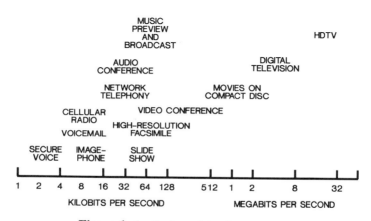

Figure 1. Applications of signal compression.

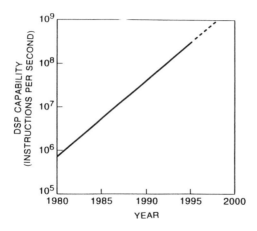

Figure 2. Evolution of arithmetic capabilities in digital signal processing.

Common to all coding algorithms for audiovisual signals are two fundamental principles (Figure 3): the removal of statistical redundancy in the input signal, and the matching of the quantization function in coding to the human perceptual system.

The redundancy-removal operation is very powerful for signals with a universal and well-understood production model, and this is exemplified by the excitation-filter model of speech production (Figure 4). In practice, the vocal-cord and vocal-tract filters that shape the voice excitation energy into the (redundant) speech waveform are modeled by so-called linear prediction algorithms. Motion compensation in interframe video coding is another important example of redundancy removal.

Principles of Signal Compression

Removal of signal redundancy

Matching of quantization to the human perceptual system

Figure 3. Principles of Signal Compression.

The matching of quantization to the human perceptual system is exemplified by the so-called JND-algorithm for perceptual audio coding (Figure 5). The just-noticeable-distortion (JND) profile in the figure defines for every frequency in the input spectrum, a critical level of distortion: as long as the distortion due to compression is at or below that critical level, the algorithm results in perfect signal quality. The JND is a function of the properties of human audition as well as the input; it is recomputed dynamically for every new block of audio, as frequently as a hundred times a second. This results in a complex encoder providing a very high level of compression. In particular, every frequency component in the input signal that falls below the JND staircase is simply discarded, with no impact on audio quality. This is how a 706 kbps CD-signal can be compressed to 64 kbps with no loss of perceived audio quality.

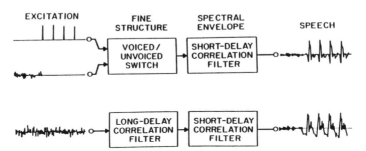

Figure 4. The excitation-filter model of speech production. After Atal [2].

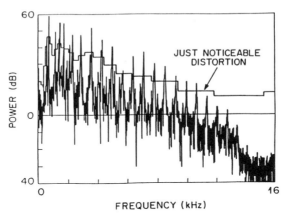

Figure 5. The just-noticeable distortion as a function of frequency for an audio signal. After Johnston [6].

The JND principle also extends to image coding (Figure 6). In this illustration, from a 2-dimensional sub-band coder, the left image is a sub-band of low horizontal and low vertical frequencies. The right image is a distortion-sensitivity profile for the specific input on the left. Dark and light areas in the right image represent areas in the original image that have low and high visual sensitivity to distortion. Gray areas represent intermediate levels of sensitivity. The sensitivity profile defines a bit allocation algorithm that neither undercodes nor overcodes the left image. By repeating this kind of critical coding for every sub-band of the input image and by synthesizing the sub-bands, a full-band compressed image is obtained.

Figure 6. Input image and the just-noticeable-distortion profile. After Safranek and Johnston [9].

Perceptual image coding leads to transparent coding of a 24-bit color image at bit rates ranging from 0.25 to 2.0 bits per sample, depending on the input picture and the viewing distance. The bit rate of 0.25 per sample has the interesting connotation of a 1-second transfer of a 512 x 512 pixel image over a 64 kbps transmission link (Figure 7).

The rate of 0.25 bit per sample is also interesting for video applications such as CD-ROM multimedia and high-definition television over a 6 MHz channel (Figure 8). In these video applications, substantial degrees of statistical and perceptual redundancy permit fairly high levels of quality at a mere quarter bit per sample. In speech and audio applications, high quality at 0.25 bit per sample is a much more difficult (and in the case of audio, perhaps unrealizable) target (Figure 9). For telephone speech and CD-audio, 0.25 bit per sample corresponds to per-second rates of 2 kbps and 11 kbps, respectively.

As we seek to enhance the current capabilities in signal compression, several generic techniques offer significant promise in preliminary research. Examples are fractal coding that utilizes self-similarities in synthetic and real images (Figures10 and 11); wavelet transforms that permit greater flexibilities in time-frequency analysis than a Fourier transform and related conventional transforms (Figure12); and highly parametrized source models in speech and particular classes of image (Figure 13).

Illustration of Perceptual Image Coding

Color Frog	24 bits per pixel
Coded Frog	1 bit per pixel
Coded Frog	0.5 bit per pixel
Coded Frog	0.25 bit per pixel

Transmission times over 64 kbps channel

100 seconds at 24 bits per pixel

1 second at 0.25 bit per pixel

Figure 7. Transmission times over 64 kbps channel.

Digital Television Formats

CIF: 360 x 288 x 30 = 3 Mpps
HDTV: 1280 x 720 x 60 = 60 Mpps

Bit Rates at 0.25 bit per pixel

CIF: 0.75 Mbps ← VIDEO DEMO (CD-ROM MULTIMEDIA)
HDTV: 15 Mbps

Figure 8. Digital Television Formats and Bit Rates at 0.25 bit per pixel.

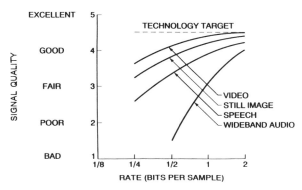

Figure 9. Current capabilities in the coding of audiovisual information.

Figure 10. A highly compressible image in fractal coding. After Barnsley [3].

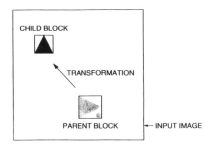

Figure 11. Coding of an arbitrary image based on soft-fractal analysis. After Jacquin [4].

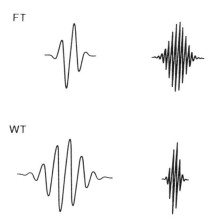

Figure 12. Qualitative comparison of Fourier and wavelet transforms. After Vetterli and Herley [11].

Complementary to signal compression are the technologies of channel coding and multi-user networking. Although a rigorous joint optimization of source coding, channel coding and networking is an intractable problem, opportunities for joint design do exist.

One example is unequal error protection in channel coding. By matching the degree of transmission error protection to the error-sensitivities of different bits at the output of a signal compression encoder, one can achieve significant gains in signal quality at the output of an imperfect transmission medium. The simplest case is a 2-level error protection algorithm where one defines two levels of error-sensitivity in the bit stream (Figure 14).

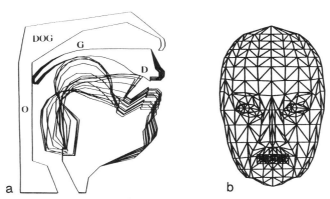

Figure 13. (a) Ariculatory model for speech coding and (b) wire-frame model for coding images of the human face. After Schroeter and Sondhi [10] and Aizawa et al [1].

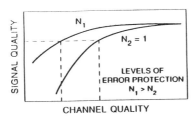

Figure 14. Unequal error protection. Providing a greater degree of protection to more sensitive bits extends the usable range of channel quality. In the simplest case of unequal protection, there are two layers of error protection (N=2).

A very similar theme is a 2-level prioritization strategy in a packet network. Signal bits are assembled separately into essential packets and enhancement packets. The guaranteed packet loss probability is designed to be much lower for the essential subset. Frequency-domain video coders such as a 2-D transform coder or a 3-D sub band coder have a built-in hierarchy that is well suited for 2-layer packetization. In general, the lower spatio-temporal frequencies tend to predominate in the essential information layer in the packet network (Figure 15).

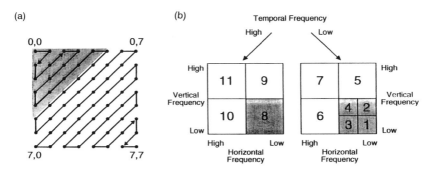

Figure 15. Two-layer coding for packet video. In the 2-D transform and 3-D sub-band coders shown, the (shaded) essential layer is guaranteed a much lower probability of packet loss.

Network theorists tend to talk about a quality of service (qos). This is a multidimensional parameter, one that is very difficult to optimize. Even in the simpler scenario of a single-user network (a source coder followed by a channel coder), there are at least four dimensions of performance: quality, efficiency, complexity and delay (Figure 16). In the coding of audiovisual signals, these parameters are measured by the mean opinion score (mos); the bit rate (bps); code complexity (mips and mW: millions of instructions per second and milliwatts); and milliseconds (ms). The delay and complexity metrics are the same in channel coding. However, efficiency is now measured in the bit rate one can squeeze into a power- and band-limited analog channel (bps/Hz), and the quality is measured as a bit error rate Pe. Systems that jointly design source and channel coding functions have the potential of maximizing signal quality at the output of a communication channel.

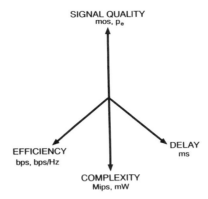

Digital Communication

Figure 16. The four dimensions of communications performance.

Researchers in signal coding have been successful in populating the four-dimensional space in our illustration with a variety of permissible scenarios, thus providing the application engineer with a menu of options with different tradeoffs along the four dimensions of performance. A formidable challenge for network theorists is to add to this knowledge base by quantifying additional dimensions of performance for digital communication. One obvious metric is the packet loss probability, which supplements the notion of bit error probability mentioned earlier. Other metrics that come to mind are the probability of access and the time for access. By definition, a good communication network seeks to minimize the bit error rate, the packet loss rate and the time for network access; and it seeks to provide a probability of access that is arbitrarily close to unity.

Joint optimization of source coding, channel coding and multi-user networking is a formidable, and generally intractable problem. Nevertheless, as we project the disciplines of signal coding and networking into the next decade, we find various opportunities for interactive design, using increasingly sophisticated models for the metrics that the customers of the communication network use to judge the overall quality of service.

REFERENCES

[1] K. Aizawa, H. Harashima and T. Saito, "Model-Based Analysis-Synthesis Image Coding (MBASIC) System for a Persons Face," *Signal Processing: Image Communication,* pp. 139-152, Elsevier Science Publishers B.V., Oct. 1989.

[2] B. S. Atal, "High-Quality Speech at Low Bit Rates: Multi-Pulse and Stochastically Excited Linear Predictive Coders," *Proc. ICASSP*, pp. 1681-1684, 1986.

[3] M. Barnsley, *Fractals Everywhere*, Academic Press, 1988.

[4] A. E. Jacquin, "A Novel Fractal Block-Coding Technique for Digital Images,"*Proc.ICASSP*, pp. 2225-2228, 1990.

[5] N. S. Jayant, "Signal Compression: Technology Targets and Research Directions," *IEEE Jour. Selected Areas in Communications,* Special Issue on Speech and Image Coding, June 1992.

[6] J. D. Johnston, "Transform Coding of Audio Signals Using Perceptual Noise Criteria," *IEEE J. Sel. Areas in Commun.,* pp. 314-323, Feb. 1988.

[7] C. I. Podilchuk and N. Farvardin, "Perceptually Based Low Bit Rate Video Coding," *Proc. ICASSP*, 1991.

[8] A. R. Reibman, "DCT-Based Embedded Coding for Packet Video," *Signal Processing: Image Communication,* pp. 333-343, September 1991.

[9] R. J. Safranek and J. D. Johnston, "A Perceptually Tuned Sub-band Image Coder with Image-Dependent Quantization and Post-Quantization Data Compression," *Proc. ICASSP,* 1989.

[10] J. Schroeter and M. M. Sondhi, "Speech Coding Based on Physiological Models of Speech Production," in *Advances in Speech Signal Processing,* ed. S. Furui and M. M. Sondhi, Marcel Dekker, New York, 1991.

[11] M. Vetterli and C. Herley, "Wavelets and Filter Banks: Theory and Design," *submitted to IEEE Trans. ASSP,* 1992.

CURRENT TRENDS IN DIGITAL VIDEO

Jules A. Bellisio

Executive Director
Video Systems and Signal Processing Research
Bellcore
Red Bank, New Jersey 07701-7040

For many years there has been a debate as to whether video would ultimately be transmitted at high bit rates using inexpensive facilities, or at low rates using powerful but low cost compression equipment. As it is developing, no debate is needed because these two technological forces are now effectively complementing each other rather than competing. Even with optical fiber where wide bandwidths are relatively easy to achieve, it turns out that when total systems considerations are analyzed, significant amounts of video compression are justified. Furthermore, optical systems are rarely used in isolation. There is frequently interworking between broadband fiber and other transmission, switching, or storage systems with more modest bit rate capability. These other systems may necessitate the use of video compression making it only logical that the fiber systems also capitalize on the benefits of the compression that must be in the system anyway.

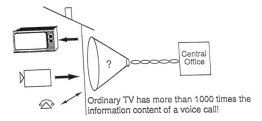

Figure 1. Video on Existing Telephone Wires.

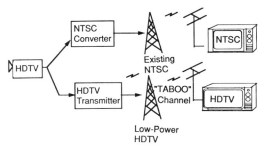

Figure 2. The Simulcast Solution.

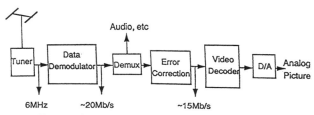

• The same basic compression technology is used from low bit rate to HDTV

Figure 3. "All Digital" Broadcast HDTV Set.

Modern video compression is making many key applications possible. We find that scaled versions and variations of the same basic technology are being used for: the transmission of video on ordinary telephone wires (64 to 1.5 Mb/s); the broadcasting of HDTV on terrestrial r-f channels (~15 Mb/s); and the storage of movies on compact discs (~700 MByte). Realizing that ordinary (uncompressed) TV has more than 1000 times the data rate of a voice phone call, it is clear that video compression technology has finally arrived and it really works. Moving up the quality scale, video compression applied to HDTV will allow broadcasters to transmit pictures via existing 6 MHz-wide terrestrial channels and thereby compete with HDTV from other sources. Digital technology has so vastly improved the working efficiency of over-the-air transmission that many designated TV channel slots, currently unusable with NTSC because of intercity interference, can now be activated and used for high-performance HDTV transmission. The next generation "all-digital" residential TV receiver will bring us not only sharper, wider pictures, but also introduce into the consumer domain very powerful signal processing engines which intrinsically can do far more than just decompress broadcast HDTV. It will bring into widespread use the technology which will enable a host of multimedia applications using a variety of delivery systems. Shown in Fig. 4 are some of the delivery alternatives and typical channel rates.

Benchmark Channel Rates for Consumer-Grade
Equipment

	Mb/s
Existing Residential Telephone Wires	1.5
Standard Compact (Audio) Disc	1.5
Terrestrial Broadcast	20
Digital VCR	32
Direct Broadcast Satellite	30
Wide-Channel CATV	45
BISDN	155

Figure 4. Some Digital Alternatives for Video.

Fitting video into an available channel involves first adjusting the size and resolution of the picture to that which is minimally acceptable, followed by the actual video data rate compression. Figure 5 illustrates resolutions used for various market segments. We note, for instance, that pictures used for 64 Kb/s videophone may have 1/16 the number of resolvable elements as normal NTSC. HDTV, on the other hand, with about twice as many lines as NTSC, each with double resolution, and a picture 1/3 wider has about 5 times as many elements which need to be coded and transmitted as NTSC.

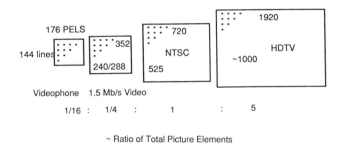

~ Ratio of Total Picture Elements

Figure 5. The Resolution Heirarchy - Approximately.

Two of the basic techniques found in nearly all video compression systems today are motion compensated interframe prediction and transform coding. When television is transmitted in its uncompressed form, each image frame is sent in its entirety and independently of all others. Because video frames, as with motion picture film, consist in large part of a series of very similar images, compression capitalizing on this temporal redundancy is possible. The modern decompressing receiver first attempts to build the current image frame by using information already available from previous frames. It does this by segmenting a previous frame into small subimage rectangles, then, by using offsetting vectors sent from the transmitter, builds a mosaic that best represents the desired current frame using these available image fragments. The vectors compensate for any movement of objects which may have occurred between the frames. Having taken this first step of predicting the current frame using compensated previous information, the decoder now receives detailed correction information which, when added at each picture element location, turns the predicted frame into the actual current frame.

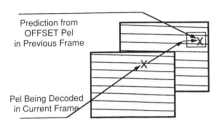

Figure 6. Motion Compensated Interframe Prediction.

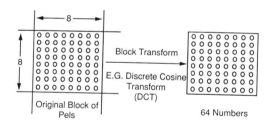

Figure 7. Transformation Coding.

Because the residual errors following motion compensated interframe prediction are themselves a frame of spatially correlated data, a transformation technique is used to further reduce bit rate. The discrete cosine transformation (DCT) using linear, reversible arithmetic converts a block of numbers representing prediction errors into another group of numbers containing equivalent information. The numbers in this second group, however, can be rank ordered in terms of their subjectively perceptible importance to the original block. In most cases, we can delete or transmit with reduced accuracy many of the lower ranked numbers following transformation. The DCT, then, is a very effective way of capturing some of the spatial redundancy of television and reducing the bit rate .

In summary, video with compression has finally arrived, it works well, and will be a dominant force across a broad spectrum of applications. The new technology has reduced the required bit rate for services that we have historically classed as "broadband." Consequently, many new services may become available using improvements of existing systems and before a full broadband infrastructure is widely deployed. This is good news, because this will build both demand for new services and traffic within the core of our networks. Full broadband deployment will become even more attractive. Digital video for wirepairs, CDs, and HDTV mark just the beginning of the digital broadband era.

PROGRESS IN FREE-SPACE DIGITAL OPTICS

H. S. Hinton

McGill University
Montreal, PQ, Canada

INTRODUCTION

Free-space digital optics is a new emerging hardware platform designed to build digital systems supporting digital signals as opposed to analog systems supporting digital signals[1]. These digital systems can optically interconnect discrete logic gates, smart pixels, and electronic multichip modules (MCM). The goal of this new platform is to complement and enhance the existing digital electronic technology with free-space interconnections.

BENEFITS OF FREE-SPACE DIGITAL OPTICS

As a result of "quantum impedance conversion," optical interconnections can provide an energy advantage in communicating logical signals from one integrated circuit (IC), multi-chip module, or printed circuit board (PCB) to another[2]. Figure 1 illustrates the required energy for an optical interconnection based on modulators compared with two versions of electrical interconnections.

A by-product of lower communication energy is that less on-chip power will be dissipated per pin-out. The importance of this is becoming evident with the push for electronic components to increase the number of pin-outs per chip. As the number of pin-outs per chip increases, one of the limiting system factors will be the power dissipated on the chip[3]. Thus, optical interconnects have the promise of extending the capabilities of systems composed of electronic ICs, MCMs, and PCBs.

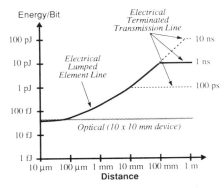

Figure 1. Minimum required communication energy as a function of distance for different bit durations.

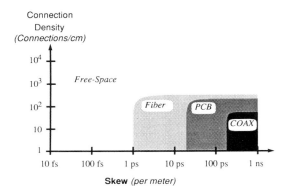

Figure 2. A comparison between electrical and optical skew and connection density.

Another benefit of optical interconnection is its low skew[3]. When building large systems that require any form of synchronization, a limiting system parameter is skew, which is the maximum difference in propagation delay between parallel channels.This skew can be caused by either dielectric or dimensional variations in the system interconnects. The skew for several interconnection technologies is shown in Figure 2.

Using this enhanced connection and communications capability of optics, new systems can be designed that maximize connections between the processing nodes (smart pixels), ICs and MCMs rather than minimizing them as in current electronic systems. This provides the opportunity to avoid the communication bottlenecks associated with connection-constrained architectures (e.g. buses), thereby leading to new high performance connection-intensive processors and switching fabrics (see Figure 3).

These new connection-intensive systems are normally composed of concatenated Two-dimensional Optoelectronic Integrated Circuits (2D-OEICs) interconnected with either bulk optics or holograms. These 2D-OEICs are a mixture of electronic and optical devices and can be thought of as electronic chips with photonic I/O. When a 2D-OEIC is composed of a two-dimensional matrix of optoelectronic circuits, it can be referred to as a smart pixel array. This unlikely mixture of electronic and optical devices is designed to take advantage of the strengths in both the electrical and photonic domain. The photonic devices include detectors to convert the signals from the previous 2D-OEIC to electronic form and either modulators, surface emitting lasers, or LEDs to transfer the results of the electronically processed information to the next 2D-OEIC. The electronics does the intelligent processing on the data, and the photonic devices provide the connectivity. Finally, high fan-out architectures could even further utilize the potential connectivity offered by the optical domain.

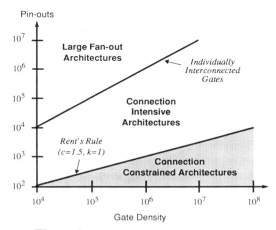

Figure 3. System architectural design space.

FREE-SPACE DIGITAL OPTICAL HARDWARE

To pursue connection-intensive architectures with large chip-to-chip connectivity, most of the proposed and demonstrated photonic interconnection networks and optical pipeline processors have been multi-stage systems. A general purpose, modular, Optical Hardware Module (OHM) is used to package the components required for each stage. A complete system would then be composed of a number of concatenated OHMs. An example of AT&T's System$_4$ OHM is shown in Figure 4[4]. This OHM is composed of four basic components; 1) active devices (SEED technology), 2) spot array generator, 3) optical interconnect, and 4) beam combiner[1].

The active devices used in AT&T's optical processor and switching fabrics have been based on the SEED technology[5]. In particular, the symmetric-SEED (S-SEED) has been the device of choice of all the AT&T systems experiments to date. It is composed of two electrically connected MQW p-i-n diodes, as illustrated in Figure 5.

To take further advantage of the spatial bandwidth available in the optical domain, electronic circuits have been integrated with optical detectors (inputs) and modulators or microlasers (outputs). This mixture of the processing capabilities of electronics and the communications capabilities of optics will allow connection-intensive architectures with

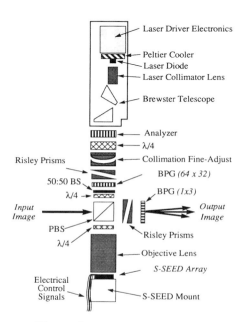

Figure 4. Optical hardware module.

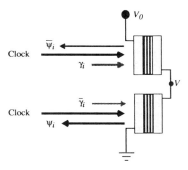

Figure 5. Symmetric-Self Electro-optic Effect Device (S-SEED)

more complex nodes to be implemented. In addition, the gain provided by the electronic devices should allow higher speed operation of the nodes. Integrated quantum well modulators with GaAs field effect transistors (FETs) have been demonstrated[6].

The second section of an OHM is the spot array generator. It provides a 2-D array of spots to clock or power the 2-D arrays of SEED-technology-based devices. These optical power supplies must split a high power laser beam into 2-D arrays of uniform intensity spots[7]. Phase gratings are one solution to this problem. These gratings are made by etching glass with a repetitive multilevel pattern. For the case of a binary phase grating (BPG), two thicknesses of glass are commonly used. This grating is illuminated by a plane wave from a laser source. The light transmitted through the grating is transformed into a 2D array of uniform intensity spots of light at the back focal plane of a lens.

The optical interconnect, the third section of the OHM, provides the chip-to-chip interconnection between the smart pixel arrays. Various forms of interconnects have been implemented using both bulk optics and holograms[7].

The final section of an OHM is the beam combiner. Free-space photonic interconnection networks or parallel processors based on SEED technology arrays require that each device be able to receive two or more input signals plus a clock signal. In addition, the reflected output signal must be directed from the same device to the next stage of the system. The major constraint is the small spots size (< 5 µm), requiring that the entering signals use the full aperture of an objective lens, and that the signals not interfere at the device's optical window.

DISCUSSION

As the bit-rates and clock rates of digital systems continue to increase it will force PCBs, MCMs, and even ICs to be capable of supporting aggregate I/O capacities in excess of 1 terabit. This is especially true in connection-intensive systems such as switching fabrics. Figure 6 illustrates the aggregate capacity as a function of connectivity (pin-outs per chip) and per channel data rate (bits per second). The capabilities of electronic modules (PCBs, MCMs, and ICs) are located in the lower left corner of the figure. The upper right is the high performance region supporting greater than a terabit aggregate capacity that can be accessed though smart pixels and free-space optical interconnects. Also included in this design space is the performance of AT&T's four system demonstrations.

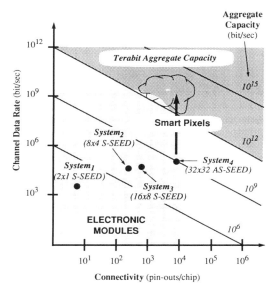

Figure 6. Device aggregate capacity.

REFERENCES

[1]. H. S. Hinton, "Photonic Switching Fabrics," *IEEE Communications Magazine,* Vol.28, No. 4, April 1990, pp. 71-89.

[2]. D. A. B. Miller, "Optics for low-energy communication inside digital processors: quantum detectors, sources, and modulators as efficient impedance converters," *Optics Letters,* Vol. 14, No. 2, January 1989.

[3]. R. A. Nordin et al, "A systems perspective on digital interconnection technology," *Journal of Lightwave Technology,* Vol. 10, No. 6, June 1992, pp. 811-827.

[4]. F. B. McCormick et al, "Digital free-space optical hardware design," *International Topical Meeting on Photonic Switching,* Minsk, Republic of Belarus, 1992.

[5]. H. S. Hinton and A. L. Lentine, "The SEED Technology," *IEEE Circuits & Devices,* To be published.

[6]. A. L. Lentine et al, "4x4 arrays of FET-SEED embedded control 2x1 switching nodes," *Topical Meeting on Smart Pixels,* Santa Barbara, CA, August 1992, Postdeadline paper.

[7]. R. L. Morrison, "Symmetries that simplify the design of spot array phase gratings," *Journal of the Optical Society of America A,* Vol. 9, No. 3, March 1992, pp. 464-471.

FDDI: CURRENT ISSUES AND FUTURE PLANS[1]

Raj Jain

Digital Equipment Corporation
550 King St. (LKG 1-2/A19)
Littleton, MA 01460-1289
Internet: Jain@Erlang.enet.DEC.Com

ABSTRACT

Key issues in upcoming FDDI standards including low-cost fiber, twisted-pair, SONET mapping, and FDDI follow-on LAN are discussed after a brief introduction to FDDI and FDDI-II.

WHAT IS FDDI?

Fiber Distributed Data Interface (FDDI) is a set of standards developed by the American National Standards Institute's (ANSI) ASC Task Group X3T9.5. This 100 Mbps local area network (LAN) uses a timed token access method to share the medium among stations. The access method is different from the traditional token access method, in that the time taken by the token to walk around the ring is accurately measured by each station and is used to determine the usability of the token.

As shown in Figure 1, older LANs, such as IEEE 802.3/Ethernet and IEEE 802.5/token ring networks, support only asynchronous traffic. FDDI adds synchronous service (see Figure 2). Synchronous traffic consists of delay-sensitive traffic such as voice packets, which need to be transmitted within a certain time interval. The asynchronous traffic consists of the data packets produced by various computer communication

---

[1] Adapted with permission from FDDI Handbook: Guide to High Speed Networking Using Fiber and Other Media by Raj Jain, to be published by Addison-Wesley, Reading, MA.

applications such as file transfer and mail. These data packets can sustain some reasonable delay and are generally throughput sensitive in the sense that higher throughput (bits or bytes per second) is more important than the time taken by the bits to travel over the network.

An important feature of FDDI, which is also reflected in its name, is its distributed nature. An attempt has been made to make all algorithms distributed in the sense that the control of the rings is not centralized. When any component fails, other components can reorganize and continue to function. This includes fault recovery, clock synchronization, token initialization, and topology control.

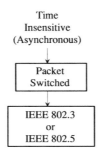

Figure 1. Service provided by IEEE 802.3 and IEEE 802.5 networks.

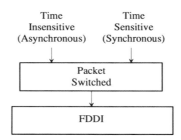

Figure 2. Services provided by FDDI.

In terms of higher layer protocols, FDDI is compatible with IEEE 802 standards such as CSMA/CD (loosely called Ethernet), token rings, and token bus. Thus, applications running on these LANs can be easily made to work over FDDI without any significant changes to upper layer software.

FDDI-II

Although the synchronous traffic service provided by FDDI guarantees a bounded delay, the delay can vary. For example, with a target token rotation time (TTRT) value of 165 ms on a ring with 10 μs latency, a station will get opportunities to transmit the synchronous traffic every 10 μs at zero load. Under heavy load, occasionally it may have to wait for 330 ms. This type of variation may not suit many constant bit rate (CBR) telecommunication applications that require a strict periodic access. For example, on an ISDN B-channel, which supports one 64-kbps voice conversation, 1 byte is received every 125 μs. Such circuit-switched traffic cannot be supported on FDDI. If an application needs guaranteed transmission of n bytes every T μs, or some integral multiples of T μs, the application is said to require isochronous service.

FDDI-II provides support for isochronous service in addition to asynchronous and synchronous service provided by FDDI as shown in Figure 3.

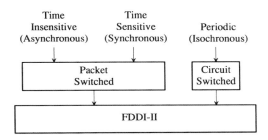

Figure 3. Services provided by FDDI-II.

Like FDDI, FDDI-II runs at 100 Mbps. FDDI-II nodes can run in FDDI or basic mode. If all stations on the ring are FDDI-II nodes, then the ring can switch to the hybrid mode in which isochronous service is provided in addition to basic mode services. However, if there is even one station on the ring that is not an FDDI-II node, the ring cannot switch to the hybrid mode and will keep running in the basic mode. In the basic mode on FDDI-II, synchronous and asynchronous traffic is transmitted in a manner identical to that on FDDI. Isochronous service is not available in the basic mode.

Most multimedia applications such as video conferencing, real-time video, and entertainment video can be supported on FDDI since the required time guarantee is a few tens of milliseconds. This can be easily guaranteed with the synchronous service and a small TTRT. Since TTRT cannot be less than the ring latency, applications requiring time bounds less than twice the ring latency cannot be supported by FDDI. Similarly, applications requiring strict periodic access will require FDDI-II. The main problem facing FDDI users is that even if only one or two stations require isochronous service, hardware on all stations on the ring would have to be upgraded to FDDI-II.

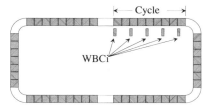

Figure 4. Cycles.

To service periodic isochronous requests, FDDI-II uses a periodic transmission policy in which transmission opportunities are repeated every 125 µs. This interval has been chosen because it matches the basic system reference frequency clock used in public telecommunications networks in North America and Europe. At this interval, a special frame called a cycle is generated. At 100 Mbps, 1562 bytes can be transmitted in 125 µs. Of these, 1560 bytes are used for the cycle and 2-bytes are used as the intercycle gap or cycle preamble. At any instant, the ring may contain several cycles as shown in Figure 4.

The bytes of the cycles are preallocated to various channels (for communication between two or more stations) on the ring. For example, a channel may have the right to use the 26th and 122nd bytes of every cycle. These bytes are reserved for the channel in the sense that if the stations owning that channel do not use it, other stations cannot use it and the bytes will be left unused.

The 1560 bytes of the cycle are divided into 16 wideband channels of 96 bytes each. Each wideband channel (WBC) provides a bandwidth of 96 bytes per 125 µs or 6.144 Mbps. This is sufficient to support one television broadcast, four high-quality stereo programs, or ninty-six telephone conversations.

Some of the 16 wideband channels may be allocated for packet mode transmissions and the others for isochronous mode transmissions. For example, channels 1, 5, and 7 may be used for packet mode transmissions or packet switching, and channels 2, 3, 4, 6, and 8 through 15 may be used for isochronous mode transmissions or circuit switching. It is possible to allocate all wideband channels for circuit switching alone or packet switching alone. The allocation is made using station management protocols, which have not yet been defined.

LOW-COST FIBER

After the initial FDDI specifications were completed in 1990, it was realized that one of the impediments to rapid deployment of FDDI was the high cost of optical components. To switch from the lower speed technology of Ethernet or token ring, it was necessary to rewire the building, install FDDI concentrators, install FDDI adapters in systems, install new software, and so on. Although the cost of all components is continuously decreasing, it is still high. Therefore, a standard effort has begun to find a low-cost alternative.

This effort has resulted in a new media-dependent physical layer (PMD) standard called Low-Cost Fiber PMD (LCF-PMD). As the name suggests, originally the committee intended to find a fiber that was cheaper than the 62.5/125 multimode fiber used in the standard FDDI. A number of alternatives such as plastic fiber and 200/230 μm fiber were considered but were quickly rejected when it was realized that the real expense was in the devices (transmitters and receivers) and not in the fiber. A search for lower powered devices then began.

LCF-PMD allows low-cost transmitter and receiver devices to be used on any FDDI link. These devices are cheaper because they have more relaxed noise margins and are either lower powered or less sensitive than those specified in the original PMD (which we prefer to call MMF-PMD, MMF stands for multimode fiber). The specification has been designed for links up to 500 m long (compared to 2 km in MMF-PMD). This distance is sufficient for most interbuilding applications.

Only the interbuilding links that are longer than 500 m need to pay the higher cost of MMF-PMD devices. Any combination of LCF, MMF, single-mode fiber (SMF), SONET, and copper links can be intermixed in a single FDDI network as long as the distance limitations of each are carefully followed.

Table 1. Low-Cost Fiber versus Multimode Fiber PMD

Issue	MMF	LCF
Wavelength	1300 nm	1300 nm
Fiber	62.5/125 multimode	62.5/125 multimode
Transmitter power	Max −20 dBm	Max −22 dBm
Receiver power	Min −31 dBm	Min −29 dBm
Connector	Duplex-FDDI	Duplex-SC or Duplex-ST
Connector keying	Port and polarity	No port. Polarity only.

Table 1 provides a comparison of the key design decisions for LCF and MMF PMDs. These are explained further below.

Wavelength

LCF uses the 1300-nm wavelength, which is the same as in multimode and single-mode PMDs. Initially, an 850-nm wavelength was suggested because 850-nm devices are used in fiber optic Ethernet (IEEE 802.3 10BASE-F) and token ring (IEEE 802.5J) networks. They are sold in large volume and so are much cheaper than 1300nm devices.

However, this would have introduced a problem of incompatibility because the users would have to remember (and label) the source wavelength and use the same wavelength device at the receiving end. This would also have caused the receiver to be replaced every time the transmitter was replaced. With 1300 nm at both ends, the user need only worry about the distance. As long as the distance is less than 500 m, the two ends can use any combination of LCF and MMF devices.

Fiber

LCF specifies 62.5/125-μm graded-index multimode fiber–the same as that specified in MMF-PMD. Initially plastic fibers and 200/230-μm step-index fibers were considered. Plastic fibers are inexpensive but they have a high attenuation. Using plastic fibers would have severely limited the distance. Two-hundred micron fibers have a larger core, which allows for a larger amount of power to be coupled in the fiber. The connectors and splices for these fibers are also cheaper since no active alignment is required. However, the large diameter of the core implies more dispersion and therefore lower bandwidth. For 200-μm fiber, a bandwidth-distance product of 30 MHz-km was predicted while 80 MHz-km (800 MHz over 100 m) has been measured. Connectors for 200-μm fiber are 30 percent cheaper; transceivers are 70 percent cheaper even while producing 10 times more power than the standard FDDI. The larger power is required because the 200-μm fiber has an attenuation of 16 dB/km compared to 2 dB/km for 62.5/125-μm fiber.

The main problem with 200/230-μm fibers is that intermixing them with 62.5/125 fibers on the same link causes a significant amount of power loss. When it was realized that a 50 percent cost reduction goal could be achieved by simply changing the transmitting and receiving power levels by 2 dBm, all efforts to change the fiber came to a halt.

Connector

The duplex connector specified in MMF-PMD was designed specifically for FDDI. Due to its low-volume production, its cost is high. Significant savings can be obtained by using other simplex connectors. In fact, many FDDI installations already use the simplex-ST connector. The LCF committee wanted to use a duplex connector to avoid the problem of misconnections (resulting in two transmitters being connected to each other). A duplex-SC connector, shown in Figure 5, was proposed. SC, which stands for subscriber connector, is a Japanese standard. It is an augmentation of the FC connector. The SC connector was developed in 1984 to provide a push-pull interface, which reduces the space required between the connectors (compared to the case if the connector has to be rotated by fingers). As a result, a large number of connectors can be placed side by side. SC has a connector loss of 0.3 dB and a return loss (reflection) of –43 dB. In the US, ST connectors are more popular than SC or FC connectors. A number of companies have proposed duplex-ST connector designs with specifications matching those of the duplex-SC. After much heated debate, termed "Connector War II," duplex-SC was voted the main selection, with duplex-ST being the recommended alternate.

Transmitter/Receivers

Reducing the transmitted power and the dynamic range (even slightly) reduces the cost significantly. LCF-PMD reduces the required transmitter power by 2 dBm and the receiver dynamic range by 2 dBm. The transmitted power range is (−22,−14) dBm while the received power range is (−29,−14) dBm. This means that the maximum loss allowed in the fiber is only 7 dB (−29,−22) instead of 11 dB. This is sufficient for a 500 m link.

Rise/Fall Times

Since LCF-PMD uses the same fiber as MMF-PMD but the link length has been decreased from 2 km to 500 m, the pulse broadening caused by fiber dispersion is less. The change then in pulse rise and fall times due to the fiber is not as much. The time thus saved has been allocated to transmitters and receivers to reduce their cost. Thus, LCF

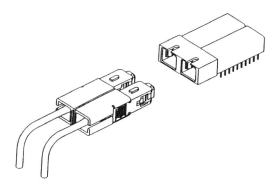

Figure 5. Duplex-SC connector

transmitters are allowed to have a rise/fall time of 4.0 ns compared to 3.5 ns for MMF transmitters. Thus, lower quality (hence, cheaper) transmitters can be used. Similarly, LCF receivers are required to receive pulses with a rise/fall time below 4.5 ns compared to 5 ns for MMF receivers. This again means less work (hence, lower cost) for the receivers.

TWISTED-PAIR PMD (TP-PMD)

As soon as the initial FDDI products started appearing on the market, the realization was made that one impediment to FDDI acceptance was that it required users to rewire their buildings with fibers. Even if you only need to connect two nearby pieces of equipment on the same floor, you will need to install fibers. Rewiring a building is a major expense and is not easy to justify unless the technology is well proven or absolutely necessary.

Besides the wiring expense, the optical components used in FDDI equipment are also very expensive compared to the electronic components used in other existing LANs. This led several manufacturers to look into the possibility of providing l00-Mbps communication on existing copper wiring. It was determined that 100 Mbps transmission using high-quality (shielded or coax) copper cables is feasible at a much lower cost than that of the fiber, particularly if the distance between nodes is limited to 100 m. Other manufacturers later found ways to transmit 100 Mbps on unshielded twisted pair (UTP), which is used in telephone wiring, up to 50 m.

An FDDI ring can have a mixture of copper and fiber links. Therefore, short links used in office areas can use existing copper wiring installed for telephones or other LAN applications. This results in considerable cost savings and quicker migration from lower speed LANs to FDDI. Proprietary coaxial cable and shielded twisted-pair (STP) products, which support FDDI links of up to 100 m, are already available. Over 98 percent of the data cable running in offices is less than 100 m and 95 percent is less than 50 m. These can be easily upgraded to run at 100 Mbps.

FDDI twisted-pair PMD is still under development. The major design issues are:

Categories of Cables

Sending a 125-Mbps signal over a coaxial cable or STP is not as challenging as on UTP. Given the preponderance of UTP cabling to the desktop in most offices, it is clear that allowing FDDI on UTP, however difficult, will be a major win for FDDI. The first issue was whether we should have different coding methods for UTP and STP or one standard covering both. A decision has been made to have one standard for both. Which categories of UTP should it cover is the next issue. While it is easier to handle data-grade twisted pair (EIA Category 5), allowing Category 3 cable would introduce more complexity.

Power Level

The attenuation (loss) of signal over copper wires increases at high frequency. To maintain a high signal-to-noise ratio, one must either increase the signal level (more power) or use special coding methods to produce lower frequency signals. Increased power results in increased interference and therefore special coding methods are required.

Electromagnetic Interference

The main problem caused by high-frequency signals over copper wires is the electromagnetic interference (EMI). After 4b/5b encoding, the FDDI signal has a bit rate of 125 Mbps. With NRZI encoding this results in a signal frequency of 62.5 MHz. At this frequency range, the copper wire acts as a broadcasting antenna. The electromagnetic radiations from the wire interfere with radio and television transmissions. The interference increases with the signal level. Federal Communications Commission (FCC) places strict limits on such electromagnetic interference (EMI). This severely limits the power that the FDDI transmitters can use, which in turn means that the distance at which the signal becomes unintelligible is also limited.

One solution to the EMI problem is to use shielded wire (STP) or coaxial cable. These wires have a special metallic shield surrounding the wires that prevents interference. Another solution is to use special coding techniques that result in a lower frequency signal. The advantage of this second approach is that the unshielded twisted-pair wires, which reach all desks, can be used for FDDI. The issue of coding has

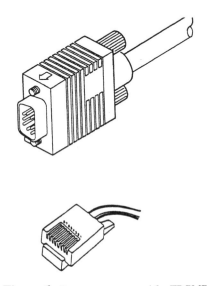

Figure 6. Connectors proposed for TP PMD.

now been resolved and a three-level coding called multilevel transmission 3 (MLT-3) has been selected. This reduces the signal frequency by a factor of 2.

Scrambling

Even though MLT-3 (and other) encoding schemes reduce the signal frequency, they are not sufficient to meet the FCC EMI requirements for UTP. One way to reduce interference is to scramble the signal so that the energy is not concentrated at one frequency. Instead, it is distributed uniformly over a range of frequencies.

Figure 6 shows the RJ-45 and DB-9 connectors proposed for TP-PMD. Both are popular connectors. They are available at a very low price due to their widespread use in computer and communication industries.

FDDI ON SONET

"SONET" stands for Synchronous Optical Network. It is a standard developed by ANSI and Exchange Carriers Standards Association (ECSA) for digital optical transmission. If you want to lease a fiber-optic line from your telephone company, it is likely to offer you a "SONET link" instead of a dark fiber link. A SONET link allows the telephone company to divide the enormous bandwidth of a dark fiber among many of its customers. Thus, a SONET link is much cheaper compared to a dark fiber link.

The SONET standard has also been adopted by CCITT. There are slight differences between the CCITT and ANSI versions. The CCITT version is called Synchronous Digital Hierarchy (SDH).

A SONET system can run at a number of predesignated data rates. These rates are specified as STS-N rates in the ANSI standard. STS-N stands for Synchronous Transport Signal level N. The lowest rate STS-1 is 51.84 Mbps. Other rates of STS-N are simply N times this rate. For example, STS-3 is 155.52 Mbps and STS-9 is 466.56 Mbps.

Table 2: SONET/SDH Signal Hierarchy

ANSI Designation	CCITT Optical Signal	CCITT Designation	Data Rate (Mbps)	Payload Rate (Mbps)
STS-1	OC-1		51.84	50.112
STS-3	OC-3	STM-1	155.52	150.336
STS-9	OC-9	STM-3	466.56	451.008
STS-12	OC-12	STM-4	622.08	601.344
STS-18	OC-18	STM-6	933.12	902.016
STS-24	OC-24	STM-8	1244.16	1202.688
STS-36	OC-36	STM-12	1866.24	1804.032
STS-48	OC-48	STM-16	2488.32	2405.376
STS-96	OC-96	STM-32	4976.64	4810.176
STS-192	OC-192	STM-64	9953.28	9620.928

Table 2 lists the complete hierarchy. The corresponding rate at the optical level is called Optical Carrier level N (OC-N). Since each bit results in one optical pulse in SONET (no 4b/5b type of coding is used), the OC-N rates are identical to STS-N rates.

For the CCITT/SDH standard, the data rates are designated STM-N (Synchronous Transport Module level N). The lowest rate STM-1 is 155.52 Mbps. Other rates are simply multiples of STM-1.

In both cases, some bandwidth is used for network overhead. The data rate available to the user, called the payload rate, is also shown in Table 2.

SONET physical-layer mapping (SPM) takes the output of the current FDDI physical layer, which is a 4b/5b encoded bit stream, and places it in appropriate bits of an STM-1 synchronous payload envelope (SPE). An STM-1 SPE consists of 2349 bytes (arranged as 9 rows of 261 bytes each. Of these, 9 bytes are used for path overhead. Since one SPE is transmitted every 125 μs, the available bandwidth is (2349x8)/125 or 139.264 Mbps. This is more than the 125 Mbps required for FDDI. The extra bits are used for network control purposes and as stuff bits for overcoming clock jitter.

SONET uses a simple NRZ encoding of bits. In this coding, a 1 is represented as high level (light on) and a 0 is represented as low level (light off). One problem with this coding is that if too many 1's (or 0's) are transmitted, the signal remains at on (or off) for a long time, resulting in a loss of bit clocking information. To solve this problem, the SONET standard requires that all bytes in a SONET signal be scrambled by a frame synchronous scrambler sequence of length 127 generated by the polynomial $1 + x^6 + x^7$. Certain overhead bytes are exempt from this requirement.

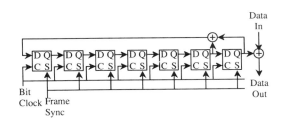

Figure 7. Shift-register implementation of a SONET scrambler.

The scrambler consists of a sequence of seven shift registers as shown in Figure 7. At the beginning of a frame, a seed value of 1111111_2 is loaded in the register. As successive bits arrive, the contents of shift registers are shifted and the sixth and seventh registers' contents (this corresponds to terms x^6 and x^7, respectively) are exclusive-or'ed and fed back to the first register (corresponding to the first term in the polynomial). The output of the final shift register is a random binary pattern, which is exclusive-or'ed to the incoming information bits.

The scrambling operation is equivalent to exclusive-oring of the bits with a particular 127-bit sequence. The sequence is highly random and does not contain long sequences of 1's or 0's. Therefore, this is expected to increase the frequency of transitions in the resulting stream. However, if the user data pattern is identical to any subset of this sequence, the resulting stream will have all 1's in the corresponding bit positions. Similarly, if the user data pattern is an exact complement of any subset of this sequence, the resulting stream will have all 0's in the corresponding bit positions.

One of the key issues in the design of the FDDI-to-SONET mapping was to ensure that the FDDI signal pattern does not result in long series of 1's or 0's after scrambling. Two steps have been taken for this purpose. First, several fixed stuff bits are used throughout the SPE to break up the FDDI stream. As a result, FDDI data cannot affect more than 17 contiguous bytes. Even the 17-byte string has one bit that is a stuff control bit; therefore, not under user control. Second, the scrambler sequence was analyzed to find the longest possible valid 4b/5b pattern that will match (or complement) a portion of the scrambler sequence. The longest possible match for random sequences of FDDI data or control symbols and the SONET scrambler sequence is 58 bits (7.25 bytes) of valid symbols. Thus, it is not possible for an FDDI user to cause serious errors in the SONET network by simply sending a data pattern.

FDDI FOLLOW-ON LAN

Both FDDI and FDDI-II run at 100 Mbps. To connect multiple FDDI networks, there is a need for a higher speed backbone network. The FDDI standards committee has realized this need and has started working on the design of the next generation of high-speed networks. The project is called FDDI Follow-On LAN (FFOL). Currently, the project is in its infancy and not much has been decided. All of the information presented here is preliminary and subject to rapid change.

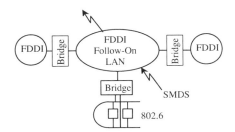

Figure 8. FDDI Follow-On network as a backbone.

The key goal of FFOL is to serve as a backbone network for multiple FDDI and FDDI-II networks. This implies that it should provide at least the packet switching and circuit switching services provided by FDDI-II. For a backbone network to be successful, it should be able to carry the traffic on a wide variety of networks. Other networks that run at speeds close to that of FDDI and that are expected to use FFOL are broadband integrated services digital networks (B-ISDN), which use asynchronous transfer mode (ATM). ATM networks use small fixed size cells. FFOL is expected to provide an ATM service that will allow the cells to be switched among ATM networks. This will allow IEEE 802.6 dual queue dual bus (DQDB) networks also to use FFOL as the backbone (see Figure 8). Easy connection to B-ISDN networks is one of the key goals of FFOL.

The key issues in the design of a high-speed network are:

Data Rate

By the time FFOL is ready, multimode fibers are expected to be in common use because of FDDI. It is desirable that users be able to use the installed fiber in FFOL. It is well known from the FDDI design that these multimode fibers have the capacity to run 100 Mbps (125 Mbps signaling rate) up to 2 km. Therefore, they can also carry a signal of 1.25 Gbps up to 200 m or 2.5 Gbps up to 100 m. The latter (100 m) covers the length of the horizontal wiring supported by ANSI/EIA/TIA 568 commercial building wiring standard. Limiting FFOL to below 2.5 Gbps will allow much of the installed multimode fiber in the buildings to be switched from FDDI to FFOL.

To carry telecommunication network traffic, FFOL should support data rates that are compatible with SONET. FFOL will be designed to be able to efficiently exchange traffic at STS-3 (155.52 Mbps), STS-12 (622.08 Mbps), STS-24 (1.24416 Gbps), and STS-48 (2.48832 Gbps).

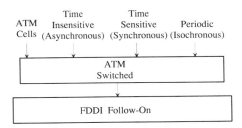

Figure 9. Services provided by the FDDI Follow-On LAN.

Media Access Modes

The term media access modes refers to the traffic switching modes supported by a network. FDDI supports three different modes of packet-switching: synchronous, asynchronous, and restricted asynchronous. Depending upon the delay and throughput requirements, an application can choose any one of these three media access modes. FDDI-II adds support for periodic (isochronous) traffic that normally requires circuit switching. FFOL is expected to support these modes. In addition, as shown in Figure 9, it is expected to explicitly support ATM switching as well. ATM switching is slightly different from packet switching. All ATM cells are the same size, the switching instants are fixed, and a slotted network design is generally used. One proposal calls for using cycles similar to those in FDDI-II and allocating some wideband channels for ATMs. Another alternative is to use an ATM base to support isochronous, ATM, and packet switching.

Media Access Method

Media access method refers to the rules for sharing the medium. Token, timed token, and slotted access are examples of the media access methods used in IEEE 802.5 token ring, FDDI, and Cambridge ring, respectively. The token access method is not useful for long distances. Its deficiency can be easily seen by considering what happens at zero load or at very high loads. Even when nothing is being transmitted, a token must be captured. This may take as long as the round trip delay around the network (ring latency). If the ring latency is D, the average access time at zero load is D/2. For networks covering large distances, this may be unacceptable. At high load, the maximum relative throughput (or efficiency) of the timed token access method is:

$$\text{Efficiency} = \frac{n(T-D)}{nT + D}$$

where, T is the target token rotation time and n is the number of active stations. The efficiency decreases as the ring latency D increases. In the extreme case of D = T, the efficiency is zero. Access methods, whose efficiency reduces with propagation delay, are also sensitive to network bit rate. Their efficiency decreases as the bit rate increases. Since the geographic extent covered by a backbone FFOL network is expected to be large, FFOL is expected to select a media access method that is relatively insensitive to the propagation delay and network bit rate.

Physical Encoding

FDDI uses a 4b/5b encoding, which allows 4 data bits to be combined into one symbol. The electronic processing is done either on symbols or on symbol pairs. These are known as symbol-wide and byte-wide implementations, respectively. Assuming a symbol-wide implementation, the electronic circuits run at 25 Mbps for 100 Mbps FDDI. At one Gbps, using the same encoding, the electronic devices will have to run at 250 Mbps. Such devices are expensive. Using larger symbol sizes such as 8/10, 16/20, or 32/40 allows parallel processing using low-speed electronic circuits. Large symbol size also allows more control symbols. These control symbols are useful for framing, fault recovery, and physical connection management.

Frame Stripping Method

Destination stripping allows spatial reuse such that the space on the media freed by the destination can be used by the destination or other succeeding nodes. Several simultaneous transmissions can be in progress in networks implementing spatial reuse. Thus, the total network throughput can be as much as n times the network bandwidth, where n is the number of simultaneous transmissions.

Destination stripping is generally used in non token rings such as register insertion rings and slotted rings. In networks using simple token access methods, multiple simultaneous transmission is not possible since each transmitting station needs a token and there is only one token. Token networks, therefore, use source stripping.

Proposals have been made for FFOL to use destination stripping. This is because the extent of the network will be large and tying up the whole medium for one transmission is not desirable.

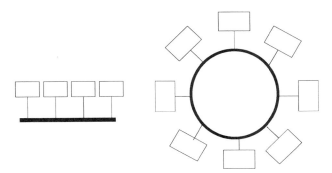

Figure 10. Shared-media distributed-switching approach.

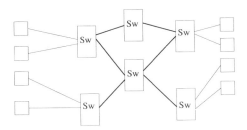

Figure 11. Shared-switching distributed-media approach.

Topology

FFOL is expected to allow the dual ring of trees physical topology that is supported by FDDI. Additional topologies may be allowed. Segments of public networks may be included in the FFOL networks. In current FDDI, only SONET links are allowed.

All LANs are designed so that the responsibility for ensuring that the packet is delivered to the correct destination is shared by all nodes. In this case, as shown in Figure 10, the switching is distributed and the medium is shared. Another alternative, shown in Figure 11, is to distribute the medium and share switches.

The advantage of this latter approach is that not all end stations need to pay the cost of a high speed connection. Their links can be upgraded to higher speeds only when necessary. The end systems are simple and most of the design complexity is in the switches. There can be several parallel transmissions at all times. Thus, the total throughput of the network is several times the bandwidth of any one link. For example, it is possible to get a total network throughput of several Gbps with all links having a bandwidth of only 100 Mbps. Notice that most telecommunication networks and wide-area computer networks use the switch-based approach. Even in high-speed LANs, there is a trend towards a switch-based mesh topology. It is not clear whether FFOL will consider a mesh topology.

Side Bar: Is FDDI a Misnomer?

The Task Group X3T9.5 was formed in 1979 to provide a high-performance I/O channel called Local Distributed Data Interface (LDDI). The idea of using optical fibers was first raised in X3T9.5 at the October 1982 meeting and subsequently LDDI standard was abandoned and a new effort based on fiber was begun. This new standard was named Fiber Distributed Data Interface or FDDI.

Initially the standard was expected to be used only on a fiber medium in a fully distributed manner for data transmission. It was expected to only specify an interface similar to SCSI (Small Computer System Interface). The features of FDDI have slowly been extended to meet diverse needs and now the name FDDI has actually become a misnomer. A more appropriate name for the current FDDI standards would be **** (four asterisks), where each asterisk stands for a wild-card in that position. FDDI is now an any-media, centralized-or-distributed, any-traffic (voice, video, or data), LAN-or-MAN-or-interface.

FDDI standards now cover non-fiber media including copper wires. Since FDDI-II uses a centralized ring master station, it is not fully distributed peer-to-peer protocol. Data was never considered to be the only traffic on FDDI. Even initial versions have features for voice, video, and other telephony applications. Finally, FDDI is a full featured network and not just another bus interface.

SUMMARY

FDDI is the next generation of high-speed networks. It is an ANSI standard that is being adopted by ISO and is being implemented all over the world. It allows communication at 100 Mbps among 500 stations distributed over a total cable distance of 100 km.

FDDI will satisfy the needs of organizations needing a higher bandwidth, a larger distance between stations, or a network spanning a greater distance than the Ethernet or IEEE 802.5 token-ring networks. It provides high reliability, high security, and noise immunity. It supports data as well as voice and video traffic.

FDDI-II provides all services provided by FDDI but adds support for isochronous traffic.

Low-cost fiber PMD allows cheaper fiber links not by using cheaper fiber but by using low-powered transceivers. The net link budget has been reduced from 11 dB to 7 dB. The reduced power allows such links to be used only if the link length is less than 500 m. Most intrabuilding links are within this distance range. The cost of the connector has also been reduced by selecting duplex versions of popular simplex connectors. These connectors are required to only have polarity keying so that an untrained user cannot misconnect a transmitter to another transmitter. No port type keying is required.

Standardization of FDDI on copper will reduce its cost considerably and help bring FDDI to the desktop.

The next higher speed version of FDDI, called FDDI Follow-On LAN, and running at 600 Mbps to 1.2 Gbps speed, is currently being discussed.

FURTHER READING

The FDDI protocols are described in a number of ANSI standards and working documents. These standards are also being adopted as ISO standards. Much of this article has been excerpted from Jain (1993). See Burr and Ross (1984), Ross and Moulton (1984), Ross (1986, 1989, and 1991), and Hawe, Graham, and Hayden (1991) for an overview of FDDI. Caves and Flatman (1986), Teener and Gvozdanovic (1989), and Ross (1991) provide an overview of FDDI-II. Stallings (1992) has a chapter devoted to the SONET standard. The analysis of the SONET scrambler for FDDI mapping is presented in Rigsbee (1990). Ginzburg, Mallard, and Newman (1990) discuss some of the problems in transmitting high bandwidth signal over copper. FFOL requirements and design considerations are summarized in Ocheltree, Horvath, and Mityko (1990) and in Ross and Fink (1992).

ACKNOWLEDGEMENT

Thanks to Bill Cronin and Paul Koning for useful feedback on an earlier version of this paper.

REFERENCES

[1] ASC X3T9.5/88-139, Fiber Distributed Data Interface (FDDI) --Media Access Control (MAC-2), Rev. 4.1 (September 3, 1992).

[2] Burr, W. E. and F. E. Ross (1984). The fiber distributed data interface: a proposal for a standard 100 mbit/s fiber optic token ring network, Proc. FOC/LAN'84, 62-65.

[3] Calvo, R. & M.Teener(1990).FDDI-II architectural and implementation examples, Proc. EFOC/LAN'90, Munich, Germany.

[4] Caves, K. and A. Flatman (1986). FDDI-II: a new standard for integrated services high speed LANs, Proc. International Conference on Wideband Communications, 81- 91.

[5] Ginzburg, S., W. Mallard, and D. Newman (1990). FDDI over unshielded twisted pairs, Proc. IEEE LCN'90, Minneapolis, MN, 395-8.

[6] Hawe, W. R., R. Graham, and P. C. Hayden (1991). Fiber distributed data interface overview, Digital Technical Journal, 3(2), 10-18.

[7] ISO 9314-1:1989, Information Processing Systems--Fiber Distributed Data Interface (FDDI) Part 1: Token Ring Physical Layer Protocol (PHY).

[8] ISO 9314-2:1989, Information Processing Systems--Fiber Distributed Data Interface (FDDI) Part 2: Token Ring Media Access Control (MAC).

[9] ISO/IEC 9314-3:1990, Information Processing Systems-Fiber Distributed Data Interface (FDDI) Part 3: Token Ring Physical Layer Medium Dependent (PMD).

[10] ISO/IEC 9314-4:199x, Information Processing Systems - Fiber Distributed Data Interface (FDDI) Part 4: Single Mode Fiber (SMF-PMD), Rev. 6.1 (May 28,1992).

[11] ISO/IEC 9314-5:1992, Information Processing Systems-Fiber Distributed Data Interface (FDDI) Part 5: Hybrid Ring Control (HRC), Rev. 6.1 (May 28, 1992).

[12] ISO/IEC 9314-6: 199x, Information Processing Systems-Fiber Distributed Data Interface (FDDI) Part 6: Token Ring Station Management (SMT), ASC-X3T9.5/84-49 Rev. 7.2 (June 25, 1992).

[13] ISO/IEC 9314-7:199x, Information Processing Systems-Fiber Distributed Data Interface (FDDI) Part 7: Token Ring Physical Layer Protocol (PHY-2),ASC-X3T9.5/88-148 Rev. 4.1 (March 5, 1991).

[14] Jain, R. (1990). Error characteristics of fiber distributed data interface (FDDI), IEEE Transactions on Communications, 38(8), 1244-1252.

[15] Jain, R., (1990). Performance analysis of FDDI token ring networks: effect of parameters and guidelines for setting TTRT, Proc. ACM SIGCOMM'90, Philadelphia, PA, 264-275. Also published in IEEE LTS Magazine, May 1991.

[16] Jain, R. (1993). FDDI Handbook: Guide to High Speed Networking Using Fiber and Other Media, Addison-Wesley, Reading, MA, Available May 1993.

[17] Ocheltree, K., S. Horvath, and G. Mityko, (1990). Requirements and design considerations for the FDDI follow-On LAN (FDDI-FO), Presentation to ASC X3T9.5 ad hoc working meeting on FDDI Follow-On LAN, FFOL-007, X3T9.5/90-068, 17 May 1990, 5 pp.

[18] Rigsbee, E. O., (1990). SONET scrambler interference analysis of ASC-X3T9.5 (FDDI) mapping data, Presentation to ASC X3T9.5 working group on FDDI/SONET Phy-layer Mapping, SPM/90-011.

[19] Ross, F. and R. Moulton (1984). FDDI overview 100 megabit per second solution, Wescon'84 conference Record, 2/1/1-9.

[20] Ross, F. (1986). FDDI-a tutorial (fiber distributed data interface), IEEE Commun. Mag., 24(5), 10-17.

[21] Ross, F. (1989). Overview of FDDI: The fiber distributed data interface, IEEE Journal on Selected Areas in communications, 7(7), 1043-1051.

[22] Ross, F. (1991). Fiber distributed data interface: an overview and update, Fiber Optics Magazine, July-August, 12-16.

[23] Ross, F. and R. Fink, (1992). Overview of FFOL - FDDI follow-On LAN, Computer Communications, 15(1), 5-10.

[24] Stallings, W., (1992). ISDN and Broadband ISDN, MacMillan, New York, 633 pp.

[25] Teener, M. and R. Gvozdanovic (1989). FDDI-II operation and architecture's, Proc. IEEE LCN'89, Minneapolis, MN, 49-61.

ROUTING PROTOCOLS RESISTANT TO SABOTAGE

Radia Perlman

Digital Equipment Corporation

ABSTRACT

A routing protocol is a distributed algorithm in which routers depend on each other to provide information in order for routes to be calculated properly and packets to be forwarded properly along those routes. In all current routing protocols a single malfunctioning router can bring down the entire network. Routing protocols claim to recover from router "failure", but the type of failure they are referring to is a "fail-stop" failure, i.e., one in which the router reverts instantaneously from working perfectly to halting. Unfortunately, not all failures are so civilized. Due to reasons such as hardware failures, implementation bugs, and misconfiguration, as well as sabotage, it is possible for a router to send misleading control information, corrupt packets, or swamp the network with too much data. This is known as a "denial of service attack".

This talk discusses a routing protocol in which data is guaranteed to be delivered provided that at least one path of correctly functioning routers and links connects the source and destination. Aside from being of theoretical interest, this protocol is of practical interest because the amount of overhead (configuration required, CPU, memory, bandwidth) is not excessively greater than that required for a traditional routing protocol.

ROBUST FLOODING

"Flooding" is a very simple routing protocol in which each router forwards a packet that it receives onto every link except the one from which it received the packet. If the network had infinite resources (bandwidth on the links, memory and CPU in the routers), flooding would guarantee that a packet would make it to the destination provided that at least one non faulty path existed from source to destination.

Unfortunately, networks do not have infinite resources. However, by making sure the finite resources are allocated fairly, we can guarantee some level of service to each source. We do this by allocating a buffer at each router for each source. To ensure that only a packet generated by the source can occupy the buffer, we use public key cryptographic signatures. To ensure that only the most recently generated packet from the source occupies the buffer, we use a sequence number.

When a router R receives a packet claiming to be generated by source S, it checks the signature to see if the packet was generated by S. If so, R checks to see if the sequence number on the received packet is greater than the sequence number of the stored packet (if any) from S. If so, R writes the received packet into memory (and deletes the old packet).

Bandwidth is allocated fairly by having R scan through its memory round robin, giving each packet a turn for transmission. As an optimization, R's neighbors can acknowledge when they receive a packet and R need no longer re-transmit a packet to a neighbor N if N has already acknowledged that packet.

In order for this to work, R needs to know the public cryptographic keys of all the sources. The method of doing this is to have a "trusted node" (or group of trusted nodes) broadcast a list of all the public keys of all the nodes in the network. This way the only configuration required is that each router needs to be configured with one public key -- the key of the trusted node. The trusted node will need to be configured with all the public keys, but that is certainly much easier than individually configuring every router with every source's public keys.

Schemes with sequence numbers must always be concerned about what happens if the sequence number space becomes exhausted, i.e., a packet is issued with the highest sequence number. Traditional schemes have the sequence number wrap around (i.e., start over again with the lowest number), or attempt to purge the network of the packets with high sequence numbers after some time, allowing the sequence number to start over again. With the robust flooding in this presentation, there is a very simple and robust solution to the sequence number space exhaustion problem. Assuming the sequence number space is sufficiently large (say 32 bits), it will only be due to malfunction of the source that high sequence numbers could ever be generated. In that case, the penalty is that the source will have to be given a new cryptographic key. Once the new key is registered with the trusted node, all the source's prior packets become invalid and the source can start over with the lowest sequence number.

ROBUST ROUTING

Robust flooding is a reasonably practical routing algorithm that has the robustness we claim. However, it is less efficient than we would like, since every packet is delivered everywhere. There are cases where that is required, as in the broadcast by the trusted node of all the sources' public keys. For data packet delivery, we would prefer if the packets traveled via a reasonably efficient path from source to destination.

We do this by using "link state routing", a traditional form of routing protocol in which each router R discovers the identity of its neighbors and broadcasts an LSP (link state packet) to all the other routers. The LSP lists the state of R's links to its neighbors. Armed with the most recently generated LSP from each other router, each router has complete knowledge of the graph and can compute routes.

We use the robust flooding described above for robust and efficient broadcast of the LSPs. Then we use "source routing", where the source router calculates a path and sets it up with a special "route setup packet". Traditional packet routing consists of each router making an independent routing decision at each hop along a packet's journey. We use source routing because it does not depend on consistent LSP databases and because routers can choose to avoid routers they are suspicious of.

The route setup packet is cryptographically protected. However, once a route is set up it is not necessary to have cryptographic protection on data packets. Instead, when S sets up a path to D, each router along the path of the route setup packet stores the direction from which it should receive packets from S to D, as well as the direction to which it should forward packets from S to D. Merely adding one check to traditional packet forwarding, which is to check that the direction from which the packet was received is the expected one, guarantees that if the source were lucky enough to choose a correctly functioning route, no malicious nodes elsewhere in the network can cause any harm.

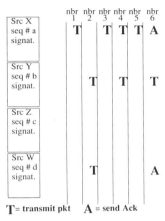

Figure 1. Database.

If the source were unlucky enough to calculate a path through a malicious node it will notice that its packets are not getting through. Then the source can choose a different path, or use flooding in desperation if a few paths don't work. In this way, assuming there are very few subverted routers (a good assumption -- a saboteur will likely take over only one router, and hardware failures causing perverse malfunctions are unlikely to occur in many routers at once), routing will be as efficient as in an ordinary network, and only when there are large numbers of malfunctions will the network go into slow but very reliable mode (falling back on flooding).

SUMMARY

The result is a robust routing protocol that is practical. Its requirements in terms of configuration, memory, CPU, and bandwidth are not significantly greater than a traditional routing scheme.

INTERNET INFRASTRUCTURE FOR PRIVACY-ENHANCED MAIL

Robert W. Shirey

Principal Scientist
Information Security Division
The MITRE Corporation

INTRODUCTION

This presentation describes Privacy-Enhanced Mail (PEM), a system to provide security for electronic messages. PEM protocols use end-to-end encryption to ensure writer-to-reader security for store-and-forward mail delivery through the Internet. This presentation also describes some of the infrastructure that will support PEM. The Internet is a rapidly growing system of several thousand interconnected computer networks that mainly support science and education. By 1992, the Internet already was serving more than a million host computers in over 100 countries. PEM was designed by a committee that is called the Privacy and Security Research Group (PSRG). A working group of the Internet Engineering Task Force (IETF) is responsible for preparing PEM specifications for publication as Internet standards.

The PSRG and IETF are both part of the Internet Society[1], an independent, international, not-for-profit, scientific and professional group that was formed in 1991.

INTERNET MAIL FLOW AND PROCEDURES

The originator of an Internet mail message uses a name and password to log in as a user at a computer that serves as a mail host. Each user has a mailbox address composed of a name for the mail host plus a name for the user at that host The user enters a command to compose a message. The host prompts for "TO" addresses, the subject, and the text. The "TO" may be either a mailbox address or the name of a locally defined

---

[1]Inquiries about the Internet Society should be sent by electronic mail to "isoc@isoc.org" or by postal mail to 1895 Preston White Drive, Suite 100, Reston, Virginia, USA 22091. The standards developed by the IETF are published in the Request for Comment (RFC) series. These documents are freely available by electronic mail or file transfer. Inquiries about the IETF and Internet standards should be sent to "ietf-info@cnri.reston.va.us".

distribution list. For an intended recipient for which the address is unknown, the user may enter a command to request the address, possibly querying a remote directory.

After writing the message, the user enters a command to send it. The host checks the format and adds fields for the "FROM" address, date, and time. The host converts each distribution list name to a list of mailbox addresses, checks for correct host names in mailbox addresses, converts host names to Internet internal addresses, and sends the message to each host mentioned in a "TO" address. If an intended destination is unavailable, the originating host stores the message and periodically attempts to send it. After a time-out period, a non-delivery notice is placed in the sender's mailbox.

STANDARDS FOR INTERNATIONAL MESSAGE HANDLING AND SECURITY TERMINOLOGY

A set of international standards, referred to as CCITT Recommendation X.400, defines the elements of a Message Handling System (MHS) that carries messages between users. An MHS includes User Agents (UAs) and a Message Transfer System (MTS). A UA is the component by which a single direct user engages in message handling. The UA is implemented by software in the user's personal workstation or in a shared host. In the Internet, the UA assists the user with the procedures described above. The Internet's UA-to-UA message format is specified by RFC 822, "Standard for the Format of ARPA Internet Text Messages." PEM is designed to be compatible with the broadest possible range of Internet UAs, operating systems, and networks. It is potentially compatible with other types of UAs, too. PEM provides security services similar to those defined in X.400, but PEM is not interoperable with X.400 security.

The MTS carries messages between UAs. The MTS consists of one or more Message Transfer Agents (MTAs) that carry messages on a store-and-forward basis. A Message Store (MS) is an optional component that is placed between an MTA and UA, and provides a capability for holding messages for later forwarding to the user. The Internet's MTA-to-MTA exchange is specified by RFC 821, "Simple Mail Transfer Protocol." PEM is intended to be implemented at or above the level of the UA-to-UA protocol, and thus be independent of MTAs.

The basic security services provided by PEM are confidentiality, origin authentication, and integrity. Confidentiality protects a message against unauthorized disclosure. Authentication provides corroboration that the source of a message received is as claimed. Integrity protects a message against unauthorized modification or destruction. If asymmetric cryptography is used for key management, PEM also provides non-repudiation of origin, which protects against an attempt by the originator to deny having sent the message. Because PEM is implemented above the UA, and because of the nature of store-and-forward message handling, certain other security concerns are not addressed. These include access control, traffic flow confidentiality, routing control, non-repudiation of destination, and message duplication, replay, or other stream-oriented problems.

CRYPTOGRAPHIC TECHNIQUES

An encryption process changes intelligible plain text into apparently random nonsense called ciphertext. A decryption process changes it back. The processes involve a mathematical algorithm and one or more variable values, called keys, that control the transformations computed by the algorithm. In symmetric cryptography, the same key is used for both encryption and decryption, and it is called the secret key. In asymmetric cryptography, a different key is needed for decryption than is used for encryption, and it is infeasible to compute one key if given the other. Therefore, one key can be made publicly available, while the other is kept private. To send a message to someone with confidentiality, we encrypt it using their public key. Only that person has the private key that is needed to decrypt the message. Some asymmetric encryption algorithms have the property that either key can be used first. With such an algorithm, we can authenticate our message by encrypting it using our private key. Anyone receiving the message can verify its origin by decrypting with our public key. The fact that we alone have the private key also provides non-repudiation of origin.

To provide message integrity, we first process the message with a message digest function, which produces a short bit string called a hash. To ensure integrity, a digest algorithm must have the property that for a given message and its hash, it is infeasible to construct another message that yields the same hash. Second, we encrypt the hash with our private key. This two-stage operation yields a digital signature for the message. Anyone receiving the message can verify its origin and integrity by recomputing the hash and comparing the result to the value obtained by decrypting the signature with our public key.

The symmetric encryption algorithms that are candidates for use in PEM are computationally more efficient than the asymmetric ones, but the symmetric ones do not have the public key property. However, we gain the advantages of both types by using a hybrid, two-step encryption process for messages. First, we select a new random number to use as a symmetric secret key to encrypt each message text. Second, we asymmetrically encrypt the secret key by using the public key of the intended recipient. The message we send includes both the symmetrically encrypted text and the asymmetrically encrypted secret key. The recipient decrypts the secret key using a private key, and then uses the secret key to decrypt the message text.

PEM uses all of the techniques just described. PEM also uses non-cryptographic transformations that put messages into canonical forms before and after encryption. These steps are necessary to ensure that a message can be correctly decrypted and read after being carried though the Internet. RFC 1421 is the current version of the message protocol. A revised draft is in progress, and a new RFC will be issued when the standard advances in the IETF.

KEY DISTRIBUTION SYSTEM

If we encrypt a message in a person's public key, we can be sure that only that person can decrypt the message. But where do we get that public key, and how can we be sure that we have the right key? The general plan is to list public keys along with their owners' names in a directory, much like telephone numbers are listed. The CCITT X.500 standard defines the components of directory database systems for computer networks. Instead of just listing the key value by itself, a directory entry will contain a public key certificate. Before directory services become widely available, users will directly exchange certificates with their correspondents.

A public key certificate is a data structure in which a public key value is cryptographically bound to a user's identity. PEM adopts the X.509 standard for defining these certificates. A certificate includes the user's name and public key, the certificate issuer's name, a serial number and a version number, a validity period, and algorithm parameters. These items are written in a standard format, hashed, and then signed with the secret key of the issuer. Once signed, a certificate can be transmitted through the Internet without additional protection. At any time, we can verify the authenticity and integrity of a certificate the same way that we check a PEM message.

The PEM standards specify procedures that anticipate the development of management hierarchies for public key certificates. For example, suppose we want to correspond with an employee in a division of a large corporation. To use PEM, the employee must already have generated a public-private key pair. The employee safely stores the private key and then takes the public key to the division's Certification Authority (CA). The CA verifies the employee's identity and then constructs and signs a certificate for the employee. When we retrieve that employee's certificate from some directory, we verify the certificate's integrity by using the public key of the CA. We get the CA's public key from the CA's certificate, which is issued by a higher level CA, say the corporate security officer. We check the division CA's certificate by using the corporate officer's public key, which comes from still another certificate at an even higher level. In other words, the certificates form a hierarchy.

PEM also adopts the X.509 standard for certificate revocation lists (CRLs). A CRL is used by a certificate issuer to notify the user community that certain certificates should no longer be considered valid even though their validity periods have not expired. A CRL lists revoked certificates by serial number. A CRL is signed by its issuer, like a certificate, and can be sent through the Internet without additional protection. There are many reasons for an issuer to invalidate a certificate. For example, a user might change employers, or a secret key might be compromised by an accident.

The PEM standard defines a tree-structured hierarchy. The standard encourages a fairly shallow tree so that the chain of computationally-intensive certificate verifications will end quickly. In the current draft plan, the Internet Society will establish an Internet Certificate Issuers Registry (ICIR), which will hold the private key that is the root of the PEM key distribution system. The ICIR's public key will be very widely published to make it easy to obtain and verify. The ICIR will issue certificates to Policy Certification Authorities (PCAs), which in turn will issue certificates to CAs. Each PCA sets the rules for the subtree beneath itself. Some PCAs are expected to require very strict controls for registering users, issuing certificates, and publishing CRLs. For example, a PCA might be established to issue certificates to CAs in some industry that uses PEM for financial transactions and requires strong controls. Other PCAs are expected to place only loose controls on their CAs, such as might be needed for electronic mail exchanged among students at a university. RFC 1422 is the current version of the protocol. A new draft is in progress, and a new RFC will be issued when the standard advances in the IETF.

ALGORITHM SUITES

RFC 1423 is the current list of PEM algorithms defined for use in PEM. For symmetric encryption, PEM uses the Data Encryption Standard as defined in Federal Information Processing Standard Publications 46 and 81, which are published in the United States by the National Institute of Standards and Technology (NIST). For asymmetric encryption, PEM uses the Rivest-Shamir-Adelman (RSA) algorithm. RSA is described in Annex C of CCITT Recommendation X.509. For hashing, PEM uses non-proprietary algorithms, MD2 and MD5, that are published by the IETF in RFCs 1319 and 1321. Additional algorithms approved for use in PEM in the future will be specified in successors to this document. Two algorithms that have already been discussed as candidates for PEM use are the Digital Signature Standard and the Secure Hash Standard, which have been proposed by NIST as new Federal standards.

DELAY AND CONGESTION IN HIGH SPEED INTEGRATED NETWORKS

Robert G. Gallager
Massachusetts Institute of Technology

ABSTRACT

The rapid evolution in optical fiber communication has made terabit link speed a likely possibility for the future. The urge to integrate voice, data, and video within a single network, combined with increasingly image oriented data traffic, also suggest that terabit networks will someday be useful. In these integrated networks of the future, different users have very different requirements. We concentrate here on delay and discuss a number of approaches to provide negotiated upper bounds on delay. We also discuss the effect that higher speeds and optical fiber technology will have on these delay bounds.

BANDWIDTH MANAGEMENT IN ATM NETWORKS USING FAST BUFFER RESERVATION

Jonathan S. Turner
Washington University

ABSTRACT

This talk describes a method for managing the allocation of bandwidth and controlling congestion in ATM networks. The method described provides a complete solution to the problem of efficient resource management in the presence of bursty traffic. The method relies on a technique called fast buffer reservation in which space in link buffers is allocated ``on-the fly" to user information bursts at the time the bursts occur, ensuring that a burst, if accepted, has extremely high probability of being delivered intact. We also describe the hardware mechanisms required to implement the method and a fast call acceptance algorithm that ensures that the probability of burst rejection is acceptably small. We conclude that the method is practical and adds only a small incremental cost to an ATM switching system. We restrict ourselves here to the method's application to point-to-point and one-to-many virtual circuits, but we note that it can also be extended to many-to-many virtual circuits (that is, virtual circuits with multiple transmitters, as well as multiple receivers).

COOPERATIVE DESIGN OVER HIGH SPEED NETWORKS

Robert E. Kahn
Corporation for National Research Initiatives

ABSTRACT

The availability of high speed networks will enable geographically separated research groups to interact on the design and implementation of complex systems. System concepts, schematic diagrams and component descriptions will need to be communicated along with plans for integration of the components or the interaction with other components. A discussion of the possibilities in this area will be given. The MOSIS system for VLSI design will be reviewed and a brief discussion of the differences between VLSI design and mechanical design will be discussed.

TRAFFIC RESEARCH FOR ATM NETWORKS

Paul J. Kühn
University of Stuttgart
Germany

ABSTRACT

The Broadband-ISDN allows the integration of arbitrary type traffic such as voice, video, image and data with constant or variable bitrates. B-ISDN is based on the Asynchronous Transfer Mode (ATM), by which small packets of constant size (cells) are multiplexed, switched and transmitted on the basis of an established virtual connection. The favorable technique is flexible to accept a variety of services differing in source traffic characteristics. The ATM technique causes, however, a set of serious traffic control and grade of service problems which have to be mastered before such networks are put into operation. Thus, worldwide activities on that subject can be observed.

The European RACE initiative aims at the development of the broadband technology as a prenormative and precompetitive joint action. Within RACE, several projects address key traffic control, modelling, performance evaluation and planning aspects, such as variable bitrate source traffic modelling, multiplexing of heterogenous cell traffic streams, finite buffer loss analysis, buffer priorities, usage parameter (rate) control, connection acceptance control, multi-stage switching network analysis and end-to-end delay jitter analysis.

The contribution reviews the major research activities in this field by identifying the problems, models, analysis approaches and results. The contribution concludes with a critical view on the state of the art and on open questions still to be attacked.

SOME ISSUES ON INTERCONNECTION OF HIGH SPEED NETWORKS

Samir Tohme
Ecole Nationale Superieure des Telecommunications, Paris

ABSTRACT

The problem of Internetworking will be very crucial in the near future between high speed networks. We are studying the best way to connect High Speed LAN to a MAN (DQDB) or an ATM Network. We extended the study to current 802.X LANs. The target was to identify the generic functions common to most of the cases, to determine the number of layers to provide assuming that we have a connectionless traffic, and how to implement these functions in the gateway. The problems of addressing and flow control have been also studied. We considered the opportunity of parallalizing some of these functions in the particular case of FDDI and DQDB. The performance evaluation of two bus architectures has been done. Numerical results about the transit time, the memory size and the error loss are available. The extension of the study to ATM is in progress. These results will be available soon.

CONCEPTS AND ISSUES IN ALL-OPTICAL NETWORKS

Charles A. Brackett
Bell Communications Research
Morristown, NJ 07962-1910

ABSTRACT

The idea of all-optical networks, in which signals traversing the network remain in optical form until they reach their destination, has become a focus of much research throughout the world. Included in this category are many kinds of time division, wavelength division, and code division multiple access network arrangements. Applications range from local area networks to national telecommunications networks and include optical interconnection within high speed information processors such as within digital switches.

This talk will review some of these developments, discuss their limitations, and address the issues which are confronting optical networks as barriers to their practicality. Particular attention will be given to the field of high-density multiwavelength networks.

COMPUTING IN THE YEAR 2020

David J. Farber
University of Pennsylvania

ABSTRACT

It is difficult to be a prophet. The field is full of such crystal ball gazers who have guessed wrong. However, even with such a caution, I will attempt to do so based on my experiences in the field and based on the quote:

> "Those who cannot remember the past are condemned to repeat it."
> George Santayana

I will describe the technological imperatives that suggest that the combination of very high speed microprocessors and gigabit plus networking will create a distributed computing environment I call "The National BackPlane". In this world, we will create over geographical distances a computing environment which mirrors in many ways the computer systems and architectures we have developed in Computer Science over the past twenty years.

Technology is not the only changing area. The use of such a National BackPlane will allow for substantial changes in the way we live as technical people and as members of our society. In the United States, Many believe we are on an Electronic Frontier where a new civilization is being formed and where changes brought about will effect the way all of us live. Terms like Cyberspace are commonly used when discussing this new environment. I wll attempt to describe the thrills and dangers of lkiving on the new Frontier and where we may be led as technological innovators and concerned humans.

CELLULAR TELEPHONE SYSTEMS: QUEUEING, BLOCKING AND GUARDBANDS

John N. Daigle
The MITRE Corporation

ABSTRACT

In this talk, we consider a service system having a finite number of servers and two classes of traffic, which arrive to the system according to independent Poisson processes. The service system has a guardband, which is a specified number of idle servers below which access will not be granted to the lower priority customers. Newly arriving lower priority customers join a finite queue if the number of servers available at the time of their arrival is less than the guardband. Meanwhile, the higher priority call requests are granted immediate service unless all servers are busy, in which case the call is dropped. We present a simple, novel approach to solving for the equilibrium probabilities for the number of lower priority calls in the queue. Additionally, we discuss blocking probabilities and other quantities of interest as a function of system parameters through the use of numerical examples.

INDEX

Adjacent channel interference (ACI) 110
ANSI 150-166
ATM 2, 8, 9-18, 26, 27-31, 36,42,109, 121, 159-166, 178, 180, 181
B-ISDN 1, 2, 9,10, 27-28, 50, 159, 180
CATV 39-41
CDMA 65-84, 85-89
Connection admission control (CAC) 1,2
Constant bit rate (CBR) 5, 19, 151
CSMA/CD 151
Database 91-96,171, 175
Discrete cosine transform (DCT) 12,137, 141
FDDI 9-10, 149-166
H.261 5,9-18,40
HDTV 15, 18, 40-42, 138-139
High Speed Digital Subscriber Line (HDSL) 40
IEEE 802 149-165
Internet 172-175
ISDN Basic rate 40
JND algorithm 128
Leaky bucket 1-8
Location registration 89, 94, 98
Location update 97-98
Markov renewal process 3
Markov chain 101-103
Message handling system (MHS) 172-173
Microcell 57, 60, 63, 65, 67, 70, 72, 83, 92, 94, 98, 108

MPEG 10-18, 40-41
Multimedia 9, 10
NIFFL 100-105
NTSC 42, 46, 138-139
OSI 110, 122
Personal Communications Network (PCN) 58, 65-80, 86-90
Personal Communications Service (PCS) 57-66
PR-NIFFL 100-104
Quality of Service (QOS) 2, 19-25, 134
Routing 24, 31, 33-35, 91-95, 112, 167-169, 174
Satellite 110-123
SEED 146-148
Segmentation and reassembly (SAR) 16
SMDS 9,10
SONET 159-166
STM 27-31
Synchronous Digital Hierarchy (SDH) 30, 158-159
TDMA 65-84, 85-89, 100, 118, 121
Traffic Burstiness 2,3,4, 8
Usage parameter control (UPC) 1
VBR 2-8, 10-18, 29
VD-NIFFL 102-106
Virtual path 2, 29, 36-37